**44** Topics in Current Chemistry
Fortschritte der chemischen Forschung

W0230492

# Cosmochemistry

Springer-Verlag
Berlin Heidelberg GmbH 1974

This series presents critical reviews of the present position and future trends in modern chemical research. It is addressed to all research and industrial chemists who wish to keep abreast of advances in their subject.

As a rule, contributions are specially commissioned. The editors and publishers will, however, always be pleased to receive suggestions and supplementary information. Papers are accepted for "Topics in Current Chemistry" in either German or English.

Any volume of the series may be purchased separately.

ISBN 978-3-662-15520-2       ISBN 978-3-540-37775-7 (eBook)
DOI 10.1007/978-3-540-37775-7

This work is subject to copyright. All rights are reserved, whether the whole or part of material is con-cerned, specifically those of translation, reprinting, re-use of illustrations, broadcasting, reproduction by photocopying machine or similar means, and storage in data banks. Under § 54 of the German Copyright Law where copies are made for other than private use, a fee is payable to the publisher, the amount of the fee to be determined by agreement with the publisher. © by Springer-Verlag Berlin Heidelberg 1974.
Originally published by Springer-Verlag Berlin Heidelberg New York in 1974
Softcover reprint of the hardcover 1st edition 1974
Library of Congress Catalog Card Number 51-5497.

The use of registered names, trademarks, etc. in this publication does not imply, even in the absence of a specific statement, that such names are exempt from the relevant protective laws and regulations and therefore free for general use.

# Contents

Interstellar Molecules

    G. Winnewisser, P. G. Mezger and H.-D. Breuer . . . . . . .     1

Carbon Chemistry of the Apollo Lunar Samples

    G. Eglinton, J. R. Maxwell and C. T. Pillinger . . . . . . . .    83

Chemistry of the Moon

    H. Wänke . . . . . . . . . . . . . . . . . . . 115

Advances in Inorganic Geochemistry

    H. Puchelt . . . . . . . . . . . . . . . . . . . 155

Editorial Board:

Prof. Dr. *A. Davison*      Department of Chemistry, Massachusetts Institute
of Technology, Cambridge, MA 02139, USA

Prof. Dr. *M. J. S. Dewar*    Department of Chemistry, The University of Texas
Austin, TX 78712, USA

Prof. Dr. *K. Hafner*       Institut für Organische Chemie der TH
D-6100 Darmstadt, Schloßgartenstraße 2

Prof. Dr. *E. Heilbronner*   Physikalisch-Chemisches Institut der Universität
CH-4000 Basel, Klingelbergstraße 80

Prof. Dr. *U. Hofmann*     Institut für Anorganische Chemie der Universität
D-6900 Heidelberg 1, Im Neuenheimer Feld 7

Prof. Dr. *J. M. Lehn*      Institut de Chimie, Université de Strasbourg, 1, rue
Blaise Pascal, B. P. 296/R8, F-67008 Strasbourg-Cedex

Prof. Dr. *K. Niedenzu*    University of Kentucky, College of Arts and Sciences
Department of Chemistry, Lexington, KY 40506, USA

Prof. Dr. *Kl. Schäfer*     Institut für Physikalische Chemie der Universität
D-6900 Heidelberg 1, Im Neuenheimer Feld 7

Prof. Dr. *G. Wittig*       Institut für Organische Chemie der Universität
D-6900 Heidelberg 1, Im Neuenheimer Feld 7

Managing Editor:

Dipl. Chem. *F. Boschke*   Springer-Verlag, D-6900 Heidelberg 1, Postfach 1780


Springer-Verlag        D-6900 Heidelberg 1  ·  Postfach 1780
Telephone (062 21) 4 91 01 · Telex 04-617 23
D-1000 Berlin 33  ·  Heidelberger Platz 3
Telephone (0 30) 82 20 11 · Telex 01-833 19


Springer-Verlag        New York, NY 10010  ·  175, Fifth Avenue
New York Inc.          Telephone 673-26 60

# Interstellar Molecules

**Gisbert Winnewisser and Peter G. Mezger**
Max-Planck-Institut für Radioastronomie, Bonn

**Hans-Dieter Breuer**
Institut für Physikalische Chemie, Universität des Saarlandes, Saarbrücken

## Contents

I. Molecules as Probes of Interstellar Space . . . . . . . . . 3
   A. What is Interstellar Space? . . . . . . . . . . . . 3
   B. Stars and Interstellar Matter . . . . . . . . . . . . 6
   C. Scope of this Review Paper . . . . . . . . . . . . 7

II. Interstellar Space and Interstellar Matter . . . . . . . . 7
   A. Introduction . . . . . . . . . . . . . . . . . . . 7
   B. Interstellar Dust and Stellar Radiation Field . . . . . . . 11
   C. Cloud Structure and Kinetic Temperature of the Interstellar
      Gas . . . . . . . . . . . . . . . . . . . . . . . 15
   D. "Dark" and "Black" Clouds and the Formation of Stars . . . 16
   E. Summary . . . . . . . . . . . . . . . . . . . . . 23

III. Observations of Interstellar Molecules . . . . . . . . . . 24
   A. Introduction . . . . . . . . . . . . . . . . . . . 24
   B. Observations . . . . . . . . . . . . . . . . . . . 26
      1. Theory of Line Emission and Absorption . . . . . . . 26
      2. Molecular Lines in the Optical and UV Range . . . . . 30
      3. Radio Observations . . . . . . . . . . . . . . . 33
   C. Identification of Interstellar Molecules . . . . . . . . 39
   D. Laboratory Spectroscopy Relevant to Astronomical
      Observations . . . . . . . . . . . . . . . . . . . 40
      1. $NH_3$-molecule . . . . . . . . . . . . . . . . . 44
      2. $H_2CO$-molecule . . . . . . . . . . . . . . . . 45
      3. $H_2O$-molecule . . . . . . . . . . . . . . . . . 48

E. Interpretation of Observed Lines . . . . . . . . . . 48
F. Observed Deviations from Local Thermodynamical
   Equilibrium . . . . . . . . . . . . . . . . . . 52
G. Conclusions and Summary . . . . . . . . . . . 55
IV.   Formation and Destruction of Molecules in Interstellar Space . . 57
A. Introduction . . . . . . . . . . . . . . . . . . 57
B. Formation of Diatomic Molecules and Radicals in the
   Gas Phase . . . . . . . . . . . . . . . . . . . 60
C. Molecule Formation on Grains . . . . . . . . . . . 63
D. Laboratory Experiments . . . . . . . . . . . . . 67
E. Lifetime and Destruction of Interstellar Molecules . . . . . 70
F. Summary . . . . . . . . . . . . . . . . . . . 72
V.   Appendix . . . . . . . . . . . . . . . . . . . . 73
VI.   References . . . . . . . . . . . . . . . . . . . 80

# I. Molecules as Probes of Interstellar Space

## A. What is Interstellar Space?

The human eye is sensitive to electromagnetic waves in the wavelength range $0.37\mu$ to $0.75\mu$. The atmosphere of the earth is transparent over a slightly wider wavelength range and a black body of temperature $\sim 4000\,°K$ radiates its maximum energy at $\lambda = 0.75\mu$; with increasing temperature this energy maximum is shifted towards shorter wavelengths. Stars, which in a first approximation may be considered as black bodies with surface temperatures ranging from $3000\,°K$ to $50000\,°K$, were for a long time the most conspicuous celestial objects and hence dominated the interest of astronomers. However, observations suggesting the existence of a circumstellar or even interstellar gas go back as far as the 17th century. Then it was the ionized part of the interstellar gas, the so-called HII regions, which attracted the interest of astronomers. An HII region is formed when a massive star of high surface temperature almost completely ionizes a large volume of the surrounding gas. This ionized gas or plasma radiates strong recombination lines in the visible range and a strong Bremsstrahlung continuum in the radio frequency range. One of the most spectacular HII regions is the Orion Nebula. Fig. 1 shows the radio contours superimposed upon an optical picture of the H$\alpha$-line emission. The contour lines are curves of constant radio emission; the black areas are regions of strong optical emission. A white wedge can be seen, extending from the East to the peak of the radio emission; this wedge has no counterpart in the radio picture. Some component of interstellar space appears to block the optical radiation without affecting the radio radiation. This component is interstellar dust; its typical grain sizes are of the same order of magnitude as the optical wavelengths, so that they attenuate light heavily yet do not affect radio waves at all. Some other examples of the attenuation of light by dust clouds are shown in Figs. 10 and 11.

The detection of interstellar gas again goes back to observations of stars. In 1904, lines of singly ionized Ca were discovered in absorption in the spectrum of the binary star $\delta$ Ori and it became clear that these lines were not associated with the stellar atmosphere but rather with the interstellar space between this star and the sun. However, three decades elapsed before the conception of interstellar matter as a gas, mostly hydrogen and helium intimately mixed with dust particles, emerged and became generally accepted.

About 1940, three interstellar radicals were detected through their UV absorption lines in stellar spectra. This made it clear that interstellar matter had another component: Interstellar molecules.

A major step forward in our knowledge of interstellar matter was made in 1951 with the discovery of the first interstellar spectral line in the radio range, the famous $\lambda 21$ cm hyperfine structure line of atomic hydrogen. Most of our present knowledge of the distribution and physical state of interstellar matter in our Galaxy is based on observations of this line. At first it looked as

3

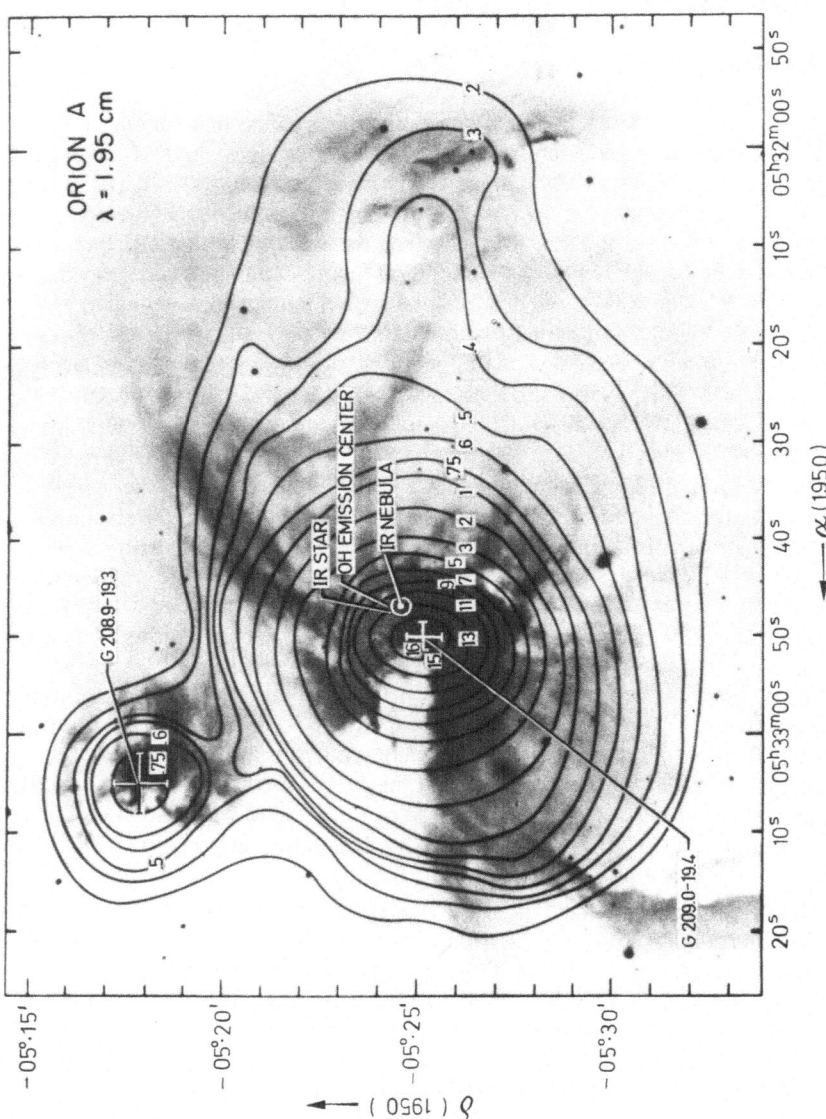

Fig. 1. Overlay of radio contours on an Hα photograph of the Orion Nebula. The difference between the radio and optical pictures is primarily due to attenuation of the optical light by interstellar dust

if the interstellar gas were rather uniformly distributed in interstellar space, with an average density of about 1 H-atom cm$^{-3}$ and a kinetic gas temperature of about 100 °K. It took another 15 years before it was realized that the distribution of interstellar gas was highly inhomogeneous in both temperature and density. There exist cool and dense clouds embedded in a hot and tenuous gas. The physical state of many of these clouds has been investigated by means of the λ21 cm line in absorption against background radio sources. In the denser clouds hydrogen exists probably in molecular form, $H_2$, and therefore such objects cannot be observed by means of the λ21 cm line. $H_2$ has no observable transitions in either the radio frequency or the optical wavelength range.

Apart from interstellar matter, the space between stars is filled with photons and cosmic particles. The UV field at wave lengths $\geqslant 912$ Å, the ionization limit for hydrogen, is the result of integrated starlight and it plays a dominant role in the formation and destruction of molecules. At wavelengths below 912 Å, most photons are absorbed by ionization of hydrogen in the immediate vicinity of the stars from which they originate. Only at wavelengths shorter than 100 Å is the ionization cross-section of hydrogen sufficiently reduced to permit photons to travel long distances through interstellar space. Fig. 2 shows a composite spectrum of interstellar UV and X-ray radiation as observed or estimated for the solar vicinity. The soft X-rays, together with the subcosmic (low-energy) part of the cosmic radiation play an important role in the heating of the interstellar

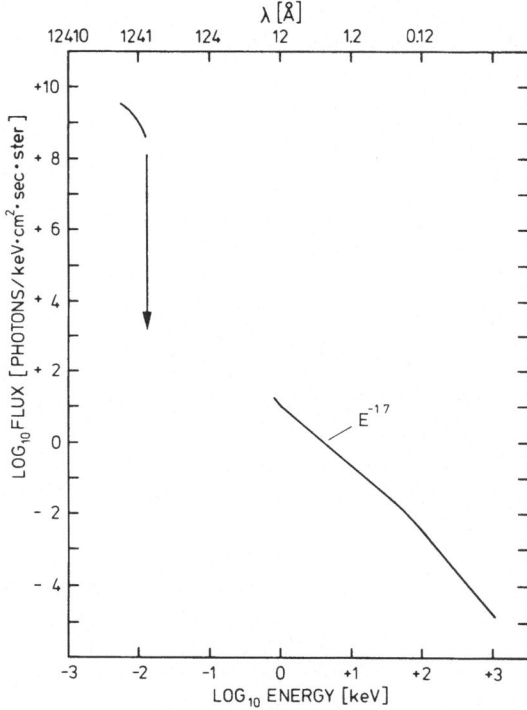

Fig. 2. Estimated interstellar UV photon flux and observed X-ray photon flux (according to Mezger, 1972). Between 13.6 eV and about 90 eV absorption of UV and X-ray photons by the interstellar gas is severe

gas, but as far as we know are of little importance for the destruction of interstellar molecules.

To summarize: the distances between stars are larger by many orders of magnitude than their diameters. The interstellar space, however, is not empty; it is filled with gas, dust particles and radiation consisting of both photons and high-energy particles. The density and temperature of the gas vary over a wide range. In the vicinity of very hot stars a large volume of interstellar gas is almost fully ionized.

## B. Stars and Interstellar Matter

The model of the evolution of the Universe that fits most of the relevant observations is the "big bang" model, which postulates that this evolution was started some $10^{10}$ years ago by expansion from a highly condensed state. In the very first minutes of this evolution hydrogen and possibly helium were formed. Later, when free electrons and protons had recombined, radiation decoupled from the expanding matter; the redshifted radiation of the primordial fireball nowadays fills the Universe as a black body radiation field with a radiation temperature of $T_{bb} = 2.7\,°K$. Its presence sets a lower limit for the temperature of interstellar material. In later stages of the expansion of the Universe, condensations formed with total masses of some $10^{12}\,M_\odot$ or less; they then contracted and, because of their inherent angular momentum, formed flat rotating gas clouds. These we observe today as stellar systems or galaxies. Stars in the mass range between 0.1 and $100\,M_\odot$ gradually condensed out of the gas. In our own Galaxy, 90% or more of the original gas has been transformed into stars. A star spends most of its lifetime in the "main sequence" burning hydrogen into helium in its core. This main sequence lifetime depends critically on the mass of the star, ranging from about $10^6$ years for a star of $50\,M_\odot$ to $10^{10}$ years for a star of slightly less than $1\,M_\odot$ The fact that we observe massive stars at present shows that stars are still forming out of the interstellar matter. Conversely we know that stars can reach their final stable "white-dwarf" stage only when their mass has decreased to less than $1.4\,M_\odot$ Various observations pertaining to mass loss from stars support this picture. Quite obviously stars and interstellar matter are not two separate entities. There exists a continuous accretion of matter into stars and replenishment of interstellar matter by mass ejection from evolving stars.

The intermediate stages between interstellar matter and the formation of protostars are seen in very dense and very cool clouds of interstellar matter. The dense clouds mentioned in the first paragraph cannot be investigated by means of the $\lambda$ 21 cm line, or any optical transition, but only by means of molecular lines in the radio frequency range. This is why interstellar molecules are so important in modern astrophysics. Molecular lines contain much information on the physical state of these dense and cool clouds. Moreover, the abundance of interstellar molecules itself can shed light on the physical conditions, including the radiation field in dense clouds, once we more fully understand the processes responsible for the formation and destruction of molecules in interstellar space.

The determination and interpretation of element abundances plays an ever-increasing role in our effort to understand the evolution of the Universe. Molecular lines may not contribute to the determination of element abundances but they already play an important role in the determination of isotopic element abundances, and these in turn can be used to discriminate between various possible thermonuclear reaction chains.

## C. Scope of this Review Paper

The following sections deal with:  II the physical conditions in interstellar space, III observations of interstellar molecules and their interpretation, including relevant laboratory measurements, and Section IV experimental and theoretical investigations of the processes of formation and destruction relevant to interstellar molecules. This review covers primarily the interstellar matter proper and refers only briefly to observations of molecules in circumstellar shells, or molecule formation in protostellar nebulae. Table 1 sets out certain quantities which are used in astronomy and are necessary to an understanding of this paper.

Table 1. Compilation of quantities commonly used in astronomy

---

1 parsec (pc) = $3.0856 \times 10^{18}$ cm = 3.2615 Light years
1 Kiloparsec (kpc) = $10^3$ pc
1 Astronomical Unit (AU) = $1.495\ 985 \times 10^{13}$ cm
1 Solar radius ($R_\odot$) = $6.9598 \times 10^{10}$ cm
1 Solar Mass ($M_\odot$) = $1.989 \times 10^{33}$ gr
Gas density $n_H$ in H atoms, expressed in terms of total density $\rho$, on the basis of observed cosmic element abundances

$$\left[ \frac{n_H}{\text{H atoms} \times \text{cm}^{-3}} \right] = 4.27 \times 10^{23} \left[ \frac{\rho}{g \cdot cm^{-3}} \right]$$

1 year = $3.155\ 6926 \times 10^7$ sec

---

## II. Interstellar Space and Interstellar Matter

### A. Introduction

Our Galaxy is one of billions of similar stellar systems which together make up the observable universe. The total mass of our galaxy is $1.8 \times 10^{11}\ M_\odot$. About 10% of that mass is in the form of interstellar matter whose principal constituents are gas and fine dust particles. The dust and gas appear to be

7

mixed, with the dust accounting for about 1% of the mass of the interstellar matter.

Table 2 indicates the chemical composition of the interstellar gas (*i.e.* its most abundant elements). Column 2 gives the abundance relative to hydrogen in number of atoms. The last four columns give the first and second ionization potentials of the element (in eV) and the equivalent wavelength of ionizing photons.

Table 2. Abundances and ionization potentials of the most abundant elements

| Element | Abundance relative to hydrogen | First ionization potential | | Second ionization potential | |
|---------|--------------------------------|----------|----------|----------|----------|
| | | eV | Å | eV | Å |
| H | 1.0 | 13.6 | 912.0 | | |
| He | $1.0 \times 10^{-1}$ | 24.6 | 504.4 | 54.4 | 227.9 |
| C | $2.3 \times 10^{-4}$ | 11.3 | 1101.2 | 24.4 | 508.6 |
| N | $1.3 \times 10^{-4}$ | 14.5 | 853.3 | 29.6 | 419.0 |
| O | $6.0 \times 10^{-4}$ | 13.6 | 910.7 | 35.1 | 353.1 |
| Ne | $2.0 \times 10^{-4}$ | 21.6 | 574.1 | 41.1 | 301.9 |
| Na | $4.4 \times 10^{-6}$ | 5.1 | 2413.0 | 47.3 | 262.2 |
| Mg | $4.4 \times 10^{-5}$ | 7.6 | 1621.9 | 15.0 | 824.8 |
| Al | $2.2 \times 10^{-6}$ | 6.0 | 2071.9 | 18.8 | 658.7 |
| Si | $4.8 \times 10^{-5}$ | 8.1 | 1521.4 | 16.3 | 758.8 |
| S | $2.1 \times 10^{-5}$ | 10.4 | 1197.1 | 23.4 | 529.8 |
| Ar | $7.1 \times 10^{-6}$ | 15.8 | 786.9 | 27.6 | 448.9 |
| Ca | $2.5 \times 10^{-6}$ | 6.1 | 2028.8 | 11.9 | 1044.7 |
| Fe | $3.2 \times 10^{-5}$ | 7.9 | 1575.4 | 16.2 | 766.3 |

Elemental abundances with exception of He are average values from Unsöld (1972, private communication). Ionization potentials as given in Allen, 1963 (Astrophysical quantities, 2nd edit. The Athlone Press, University of London 1963).

Hydrogen is the most abundant element (70% by mass), followed by helium (28% by mass). Elements heavier than helium account for the remaining 2%. The available evidence indicates that hydrogen and helium were created during the earliest evolutionary phases of the universe (*i.e.* in the "big bang") whereas the heavier elements may have been synthesized more slowly by thermonuclear reactions in stars and somehow returned to the interstellar medium. Known phenomena involving extensive mass loss from stars include supernovae, novae, and planetary nebulae. However, the details of the processes which led to the present chemical composition of the Galaxy (and of the Universe) are far from being understood. The composition of the interstellar dust is unknown. Molecular ice, silicates, and graphite have been suggested.

The stars are concentrated to a flat disk of diameter about 40 kpc and width about 500 pc, and to a spherical bulge of diameter about 2 kpc in the central region (see Fig. 3). The plane of symmetry of the disk is called the

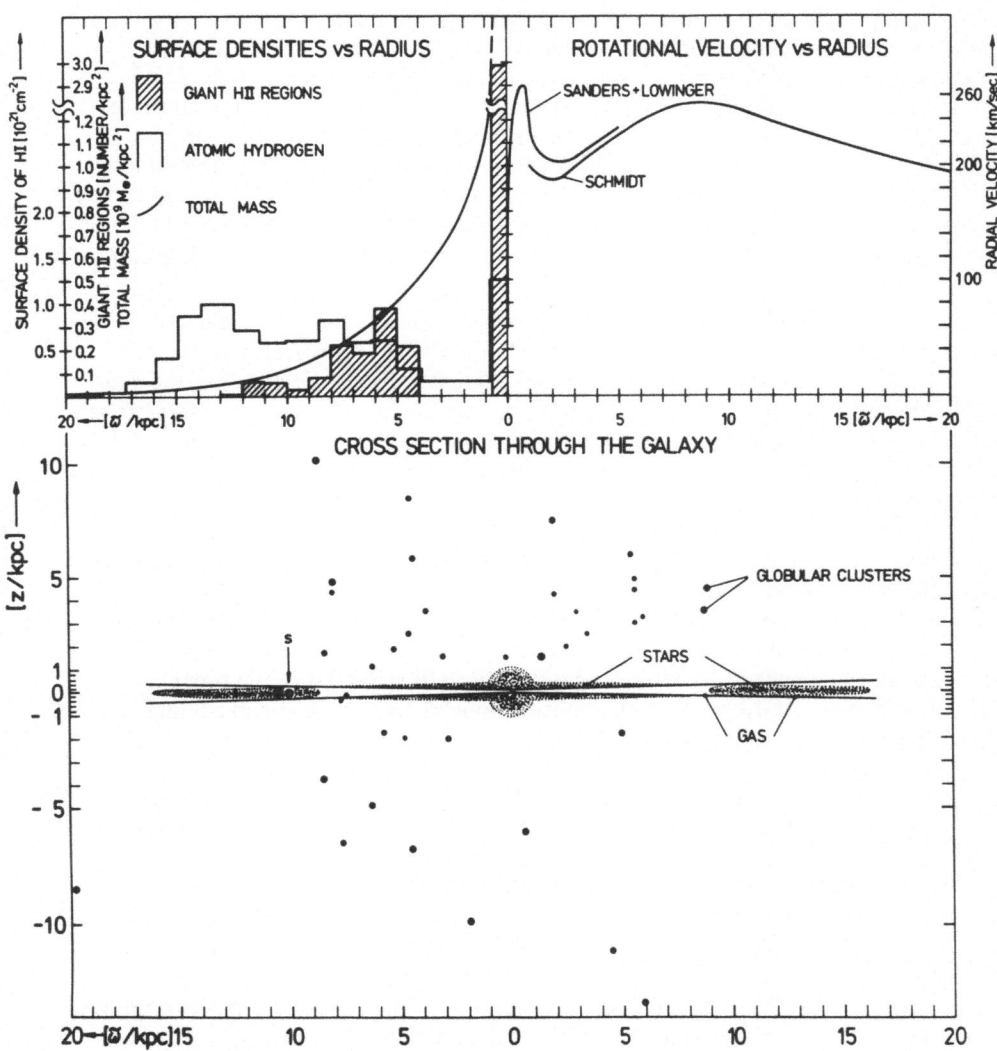

Fig. 3. Distribution of stars and interstellar matter in the Galaxy. The lower part of the diagram is a cross section perpendicular to the galactic plane. Globular clusters are the oldest stellar systems and must therefore have been formed in the early evolutionary stages of the Galaxy. The mass of the stars, however, forms a flat layer with a "nuclear bulge" in the center. The interstellar matter forms an even flatter layer which widens up towards the edges of the Galaxy.

The upper left part of the diagram shows the column density of stars, hydrogen gas and HII regions as a function of the distance $\omega$ from the galactic center. Stars and interstellar matter rotate around the center of gravity of the Galaxy. The rotational (orbital) velocity is given as a function of $\omega$ on the right of the upper diagram. At the distance of the sun this velocity is 250 km/sec; it takes the sun $2.5 \times 10^8$ yr to complete one rotation around the galactic center

9

*"galactic plane"*. The interstellar material is more closely confined to the galactic plane than are the stars. Within 4 kpc of the galactic center, the half-density width of the gas layer is about 100 pc. Between galactic radii 4 and 10 kpc, the width increases to 200 pc, and between 10 and 20 kpc it increases further to about 600 pc. Whereas the highest density of stars occurs in the central part of the Galaxy, the bulk of the interstellar gas lies between galactic radii 4 and 16 kpc. Therefore, in the vicinity of the sun at galactic radius 10 kpc and close to the galactic plane, the average proportion of gas to stars is considerably higher than the overall value of 10%; observations indicate that it is about 30%.

Within 4 kpc of the center, the average gas density close to the galactic plane is $n_H \approx 0.3$ cm$^{-3}$, increasing to about 0.7 cm$^{-3}$ between galactic radii 4 and 10 kpc and then decreasing again at greater distances. However, the density and temperature of the interstellar gas are far from uniform; local densities as high as $10^7$ cm$^{-3}$ are known and higher densities are suspected. It is probably in these small regions of high density that complex polyatomic molecules are formed.

The highly flattened disk of the Galaxy is due to the fact that the entire system is rotating. The angular velocity decreases with increasing distance from the galactic center. From the center to a distance of about 800 pc there is a very flat and rapidly rotating gas cloud which contains several $10^7\,M_\odot$; this is called the "nuclear disk". The gas in the central region out to a distance of about 4 kpc appears to be flowing outwards with velocities of the order of 100 km/sec. At 4 kpc, the radial motion appears to be braked and a piling-up of gas forms an expanding (at about 50 km/sec) ring-shaped feature referred to as the "3-kpc" arm.

Outside the "3-kpc" arm, the spiral arms are the dominant large-scale structure. Most of our present knowledge concerning the large-scale distribution of the interstellar gas in our Galaxy is based on observations of the λ21 cm hyperfine structure line of the hydrogen atom. Extensive surveys of the λ21 cm emission led to the large-scale distribution of interstellar gas as shown in Fig. 4. This distribution is certainly not a clear-cut spiral pattern, but rather consists of a number of extended features of increased gas density, referred to as "material" and "density wave" spiral arms. The gas density in these spiral arms may be as much as ten times higher than the density of the gas in inter-arm regions. The observations also reveal large variations in the kinetic gas temperature, from values as high as several $10^3$ °K in the interarm regions to values considerably below $10^2$ °K in some regions of higher density in the spiral arms.

The principal features of the spiral structure are believed to be maintained by a pattern of density waves. This pattern rotates about the galactic center like a rigid body with an angular velocity that is lower than that of most of the material. Thus, the material of the Galaxy (*i.e.* both stars and gas) flows through the pattern and as it enters the density wave, the gas is strongly compressed. There are also less important elements of the spiral pattern that rotate with the material; these are called "material" arms, as against "density-wave"

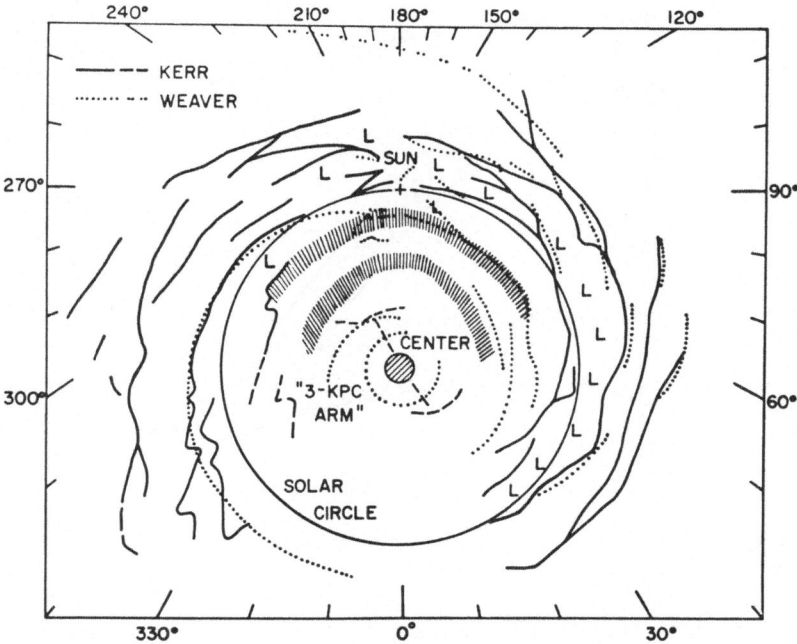

Fig. 4. Superimposed maps of the neutral hydrogen distribution in the Galaxy as interpreted by Kerr and by Weaver. The points marked "L" denote regions that Kerr interprets as markedly deficient in neutral hydrogen. Dashed lines and hatched areas indicate where the location is uncertain, not regions where the hydrogen is weak or strong. The radius of the solar circle is 10 kpc. Galactic longitude is shown around the edge of the map

arms. Fig. 5 shows a very schematic representation of the density structure (*i.e.* width and separation of spiral arms) in the vicinity of the sun. P and S are thought to be density-wave spiral arms, O appears to be a material arm. The sun is located near the edge of the arm O. It may be noted that this picture of the spiral structure in the vicinity of the sun is based on both optical and radio observations, whereas the large scale structure of the interstellar gas (Fig. 4) is based only on observations of the λ21 cm hydrogen line and kinematic distances, which become highly uncertain towards the galactic center and anticenter. Kinematic distances are obtained from a model for the rotation of the Galaxy due to Schmidt. This model relates the circular orbital velocity of the gas to the distance from the galactic center (see Fig. 3).

## B. Interstellar Dust and Stellar Radiation Field

Interstellar space is permeated by radiation of all wavelengths originating from the stars. However, below 912 Å, which is the ionization potential (Lyman

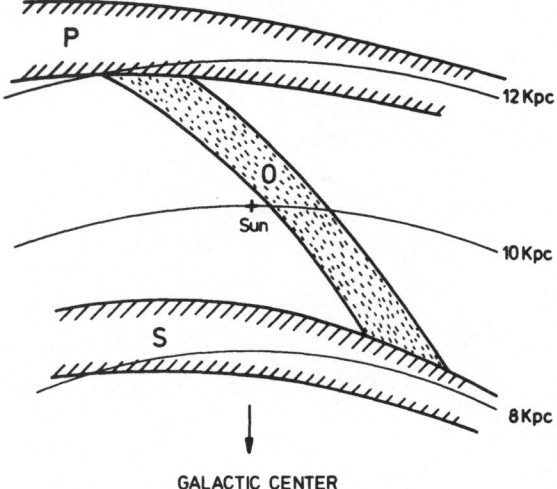

Fig. 5. Schematic representation of the spiral structure in the vicinity of the sun. S and P refer to density wave spiral arms, O to a material arm

continuum limit) of hydrogen, photons can ionize hydrogen and are therefore usually absorbed in the immediate vicinity of the star from which they originate. At wavelengths longer than the Lyman continuum limit, the principal source of attenuation is the interstellar dust. Dust grains both absorb and scatter light; the combined effect is called extinction. Fig. 6a shows a typical extinction curve.

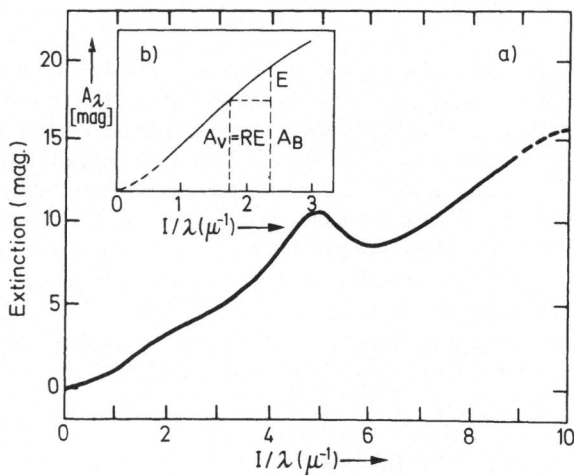

Figs. 6a und b. a) Typical observed interstellar extinction curve from the IR through the optical range to the UV. b) Schematic representation of the optical part of the extinction curve, defining the quantities $A_v$, $A_b$ and $R$

It increases rapidly with decreasing wavelength. The extinction is measured in magnitudes. From optical observations of stars, it is easy to determine the relative extinction or color excess $E = A_B - A_V$, with $A_B$ and $A_V$ the total extinction at wavelengths $B$ (= 4400 Å) and $V$ (= 5500 Å). On the assumption that the extinction curve is similar in all directions in space, the total extinction in the visual is (Fig. 6b)

$$A_V = R E \quad \text{with } R = 3.0 \pm 0.3 \tag{1}$$

determined from observations. The optical depth $\tau$ is related to the extinction $A_\lambda$ at any wavelength $\lambda$ by

$$\tau_\lambda = \frac{1}{1.086} \left[ \frac{A_\lambda}{\text{mag}} \right] \tag{2}$$

A plane wave propagating through a dust layer of optical depth $\tau_\lambda$ is consequently attenuated by the amount $\exp \{-\tau_\lambda\}$. The shape of the extinction curve in Fig. 6a can be explained by assuming a mixture of grains with a wide range of grain sizes. Most effects observed at optical wavelengths can be explained by dielectric grains of size $a \simeq 0.15\mu$. To explain the extinction in the far UV, one needs typical grain sizes between $0.02\mu$ and $0.05\mu$, depending on the material.

The albedo of a dust grain is defined as the ratio of its scattering cross section to its cross section for total extinction. From the observations, values of the albedo as high as 0.9 at $\lambda \simeq 1500$ Å are obtained. A particle with no

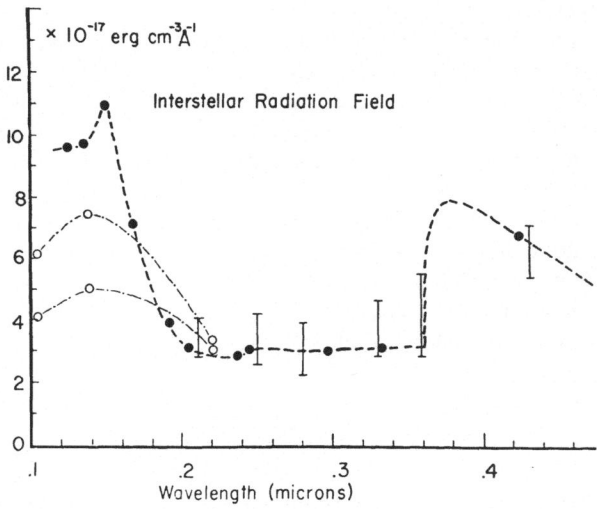

Fig. 7. Interstellar UV radiation field in the vicinity of the sun, estimated by Witt and Johnson (1972) on the basis of the albedo of dust grains determined from OAO-2 observations

dielectric losses does not absorb radiation but only scatters it, and the albedo is, therefore 1.0. However, most materials do not behave like a pure dielectric in the far UV but absorb strongly. Hence a value of the albedo of about 0.5 is expected in contradiction to the observations. If most of the extinction is due to scattering, the effective optical depth of the dust for calculating the diffuse radiation field is much smaller than the value given by Eq. (2). In fact, Watson and Salpeter (1972a, b) use, for the attenuation of UV radiation by a dust layer of optical depth $\tau_\nu = 0.92\,A_\nu$, the approximation

$$I_{UV} = I_{o,UV}\, e^{-2.5\,\tau_\nu} \tag{3}$$

independent of wavelength, with $I_{o,UV}$ the UV flux density at the surface of the cloud. Densities of the UV flux in unprotected regions of interstellar space, as estimated by Habing (1968), are given in Table 3. More recent estimates of the interstellar UV radiation by Witt and Johnson (1972) are given in Fig. 7. Note especially the increased UV flux densities shorter than 2200 Å, which are the result of the high albedo of interstellar dust grains in this wavelength range.

Table 3. Density of interstellar UV radiation according to Habing (1968)

| $\lambda/\text{Å}$ | 1000 | 1400 | 2200 |
|---|---|---|---|
| $u_\lambda \times 10^{20}/\text{erg}/(\text{cm}^3\,\text{Å})$ | 4000 | 5000 | 3000 |

Most molecules observed to date in interstellar space can be dissociated by UV radiation of wave lengths longer than 912 Å. In fact, their average lifetimes in interstellar space are $< 100$ yr, unless they are protected by a dust layer (Section IV. E). This, and the fact that surface reactions on dust grains play an important role in the formation of interstellar molecules (see Section IV. C), explains at least qualitatively the association of dense dust clouds and molecules in interstellar space.

If gas and dust are homogeneously mixed, there must be a universal relation between the column density of hydrogen $N_H$ (*i.e.* the number of hydrogen atoms per cm$^2$ integrated along the line of sight) and the corresponding extinction $A_\nu$ measured in magnitudes caused by the dust grains. The best determination to date of this relationship is that of Savage and Jenkins (1972):

$$\left[\frac{N_H}{\text{atoms cm}^{-2}}\right] = 1.7 \times 10^{21} \left[\frac{A_\nu}{\text{mag}}\right] \tag{4}$$

Two hypotheses about the origin of interstellar dust seem to be under consideration at the moment. The first assumes that grains condense in interstellar space, and the second that they condense in stellar atmospheres and envelopes. Another hypothesis combines these two, suggesting that condensation nuclei form in circumstellar nebulae and the bulk of the material of the grains consists in the form of more volatile species that could condense in interstellar space. Arguments in favor of the second hypothesis are given by Woolf (1972) and are based on studies of infrared emission and absorption of dust. Observations of the Orion nebula indicate the presence of silicates, but there is no evidence of iron or carbon. Furthermore observations of the galactic center and the Orion nebula show that the solid core of the grains may be covered by a mantle of ice whose mass is about 10% that of the silicate mass.

## C. Cloud Structure and Kinetic Temperature of the Interstellar Gas

In the interarm region (see Figs. 4 and 5 of the large scale spiral structure of the interstellar gas), one observes a hot and tenuous gas in which are embedded small condensations of relatively dense and cool gas, referred to as "cloudlets". Observations to date place rather wide limits on the kinetic temperature of the hot gas, viz. $700 \,°K \leqslant T_k \leqslant 6300 \,°K$. The average density of the hot gas appears to be low, of the order of 0.2 to 0.02 atoms $cm^{-3}$. The cloudlets have typically a density of some 10 atoms $cm^{-3}$, a total mass of some $10 \, M_\odot$ and a kinetic temperature of less than $100 \,°K$. These cloudlets, which have an average extinction of $< A_v > = 0.26^m$ per cloud, appear to be responsible for the average optical extinction of about 1 visual mag per kpc of the interstellar gas. In these cloudlets the OH $\lambda 18$ cm and the $H_2CO$ $\lambda 6$ cm lines may be observed in absorption against continuum radio sources, and the Lyman resonance bands of molecular hydrogen are seen in absorption against hot stars.

For interstellar gas in an equilibrium state, the kinetic temperature is determined by the condition that the energy input, which is primarily due to ionization, is equal to the energy loss, which occurs primarily through inelastic collisions, i.e. collisional excitation of atoms and ions resulting in the escape of radiation. The most important collisionally excited lines are Lyman $\alpha$ at high temperatures, and CII $\lambda 156\mu$ and CI $\lambda 610\mu$ at low temperatures. Since the interstellar gas in the interarm region has temperatures above $1000 \,°K$, the heating of the gas must be due to relatively energetic particles or photons such as subcosmic rays or soft X-rays. Once the ionization rate $\xi$ is fixed, the temperature of the gas depends only on the density. Because the heating of the gas is proportional to the gas density while the cooling by collisional excitation is proportional to the square of the density, the temperature of the gas decreases with increasing gas density (Fig. 8a). The product of the temperature $T$ and the particle density $n$ is proportional to the pressure (Fig. 8b). We see from this curve that a hot and tenuous gas ($T \simeq 10^4 \,°K$, $n \simeq 10^{-1} \, cm^{-3}$) can be in pressure equilibrium with cool and dense condensations ($T < 10^2 \,°K$, $n > 10 \, cm^{-3}$). This model explains qualitatively the coexistence of dense, cool gas cloudlets in the hot interarm gas. It also predicts the degree of ionization in each component. The predicted density of free electrons is given in Fig. 8b. Note that the two-component model of cool and dense condensations embedded in a hot and tenuous gas appears to reflect a general characteristic of the interstellar gas that is not related to its large-scale structure such as spiral arms.

15

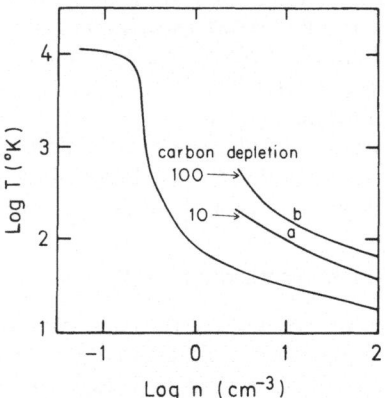

Fig. 8a. Equilibrium temperature of a gas heated by cosmic rays, ionization rate $\xi =$ $4 \times 10^{-16}$ sec$^{-1}$. The temperature rise at about 0.3 cm$^{-3}$ is associated with the inability of a low-density gas to cool rapidly by fine structure transitions; the flattening below 0.2 cm$^{-3}$ is due to Lyman-$\alpha$ cooling. If various ions effective in cooling (such as carbon) are locked up in grains, the temperature of the gas in clouds ($\curlywedge$ 10 cm$^{-3}$) will be higher

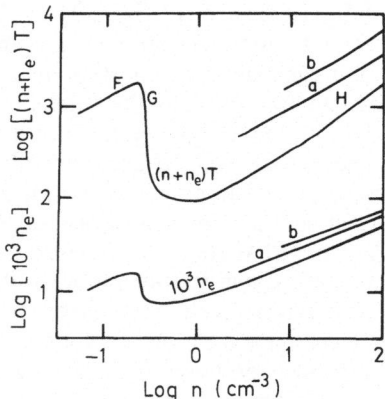

Fig. 8b. Equilibrium pressure $(n + n_e)T$ and electron density. The existence of two stable regions $(dp/d\rho > 0)$ permits two phases, F and H, to exist in pressure equilibrium. Phase F is identified with the intercloud medium, and H, with clouds. Carbon depletion lowers the density and increases the temperature of the cloud gas. The electron density in clouds is about 10 times the classical value of HI regions, $4 \times 10^{-4}$ $n_H$. The value in the intercloud medium is about $2 \times 10^{-2}$ cm$^{-3}$

## D. "Dark" and "Black" Clouds and the Formation of Stars

Observations of the interstellar gas show the existence of another type of cloud, much denser and more massive than the cloudlets. These clouds appear to be related to the large-scale structure of the interstellar gas, since they are located mainly in spiral arms and in the nuclear disk. The density and total mass of these clouds are so high that subcosmic particles and soft X-rays will be absorbed in the outer layers. Thus, the thermal balance of these clouds is not the same as that in the cloudlets. Fig. 9 shows a qualitative picture of the

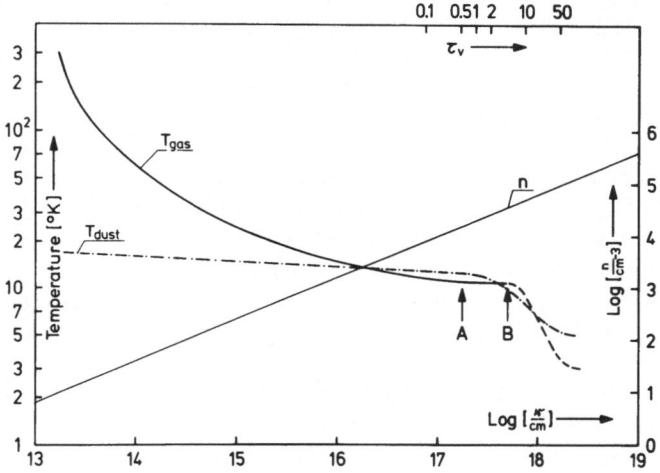

Fig. 9. Estimated change of kinetic gas temperature and equilibrium temperature of dust grains as a function of the distance $r$ to the surface of the cloud. The calculation assumes a linear change of density $n \propto r$, a constant ionization rate by subcosmic particles and penetration of UV photons $\lambda > 912$ Å into the cloud only from the outside. Absorption of subcosmic particles is neglected. The upper scale indicates the optical depth of the dust grains in the visual

change of kinetic gas temperature in a gas cloud as a function of distance from the surface of the cloud. We have assumed a linear increase in density towards the center of the cloud but have neglected attenuation of the subcosmic radiation. Thus, the decrease in temperature is primarily due to the increase in density. In the outer layers the kinetic gas temperature is still determined by the heat input provided by ionizing subcosmic particles. At point A, however, this heat input becomes negligible compared to heat input by photons in the wavelength range 912 to 1100 Å which can ionize carbon. This mechanism provides an equilibrium temperature of about 12 °K, independent of the density. At point B, all photons that can ionize carbon are absorbed and the CII region ends. The temperature then drops to a value limited by the 3 °K black body background radiation and is controlled by the temperature of the grains and the ionization by high energy cosmic rays on the one hand, and saturation of the cooling transitions on the other.

The temperature of dust grains is a measure of the equilibrium between heating by absorption of stellar light and re-emission in the far IR. In free interstellar space, dielectric grains attain a temperature of about 17 °K. With increasing optical depth of the gas layer, the heating of the dust grains and consequently their temperature decreases. This decrease is very slow. Only at an optical depth of dust grains $\tau_v \simeq 50$ (in the visual) will the grain temperature drop to values of the order of 5 °K (Werner and Salpeter, 1970). Beyond point B, it is not quite clear whether dust grains will provide the dominant heating or cooling mechanism of the gas. The minimum kinetic gas temperature of about 2.7 °K is determined by the microwave background radiation.

As long as so little is known about the interstellar particle and photon radiation fields which provide the heating of the interstellar gas, one should not put too much faith in estimated equilibrium temperatures of dense gas clouds such as shown in Fig. 9. However, at least qualitatively it appears to be correct that dense condensations in the interstellar gas have low kinetic temperatures. In fact, gas-kinetic temperatures as low as 5 °K have been observationally determined in the center of dark clouds.

Whenever the gravitational forces in a condensation become large enough to overcome the internal pressure of the gas, the condensation will begin to contract. The superiority of the gravitational forces is thereby increased and the contraction must eventually lead to the formation of stars in which, once thermonuclear reactions have started in their cores, the gravitational forces are balanced by gas and radiation pressure. Gravitational contraction will start once the mass of a gaseous condensation of hydrogen density $n_H$ and kinetic temperature $T_k$ exceeds a critical value, known as the Jeans mass

$$\frac{M_{Jeans}}{M_\odot} = 32 \, \frac{T_k^{3/2}}{n_H^{1/2}}. \tag{5}$$

In interstellar clouds of high density and low temperature, stellar masses can, according to Eq. (5), become gravitationally unstable and contract. Obviously, these clouds must represent the first stages in the evolutionary path which leads from condensations in the interstellar gas to main sequence stars. Computations show that in the denser interstellar clouds most of the atomic hydrogen is tied up in $H_2$ molecules (Hollenbach et al., 1971). It is therefore impossible to investigate the physical state of such a cloud by means of $\lambda 21$-cm line observations. Molecular lines, on the other hand, are excellent indicators for the very cool and dense clouds. It is one of the most fascinating aspects of molecular spectroscopy that it provides, at last, an observational method for the investigation of objects which lie along the evolutionary path leading from a cloud of interstellar gas to a star cluster.

Within a distance of 1 kpc from the sun, dense interstellar clouds can be observed optically because they attenuate the light of stars located behind them. Fig. 10 shows as an example the dust clouds Barnard 1968 and 1972 which are seen as dark areas in an otherwise rather uniform field of stars. Fig. 11 shows another dust cloud, the "horsehead nebula", silhouetted against the bright background of an emission nebula (HII region). Most of the extended dust clouds

Fig. 10. The dust clouds Barnard 68 and 72, photographed by B. J. Bok (private communication: at the 90-inch reflector of the Steward Observatory, Tucson, Arizona. These two dense clouds composed of dust and gas are seen projected against the great star cloud in Sagittarius

are not as conspicuous as these objects and their extinction is obtained by laborious star counts. Fig. 12 shows, as an example, a contour map representation of the Taurus cloud, located in the direction of the anticenter and at a distance from the sun of about 100 pc (Mezger, 1971). The contour lines are lines of constant extinction in visual magnitudes. We see that the Taurus cloud extends over several tens of degrees, with an average extinction[a] between $1^m$ and $2^m$.

Fig. 11. The "horsehead" nebula, another dust cloud, seen projected against an HII region, *i.e.* an extended ionized cloud of interstellar gas

There are smaller regions in this cloud (indicated in the diagram as hatched areas) where the extinction attains values of $5^m$ to $8^m$. These are the "dark clouds" proper shown in Fig. 10 and 11 and referred to in the reports of many radio-astronomical investigations. In these dark clouds, the two central OH $\lambda$ 18 cm lines are observed in quasi-thermal emission, the CO $\lambda$ 2.6 mm line is, seen in probably Local Thermodynamic Equilibrium (LTE) emission and the $H_2CO$ $\lambda$ 6 cm line is seen in anomalous absorption against the 3 °K background radiation. Interpretation of the emission lines indicates that densities in dark clouds are of the order of $10^4$ particles cm$^{-3}$ and kinetic gas temperatures are as low as 5 °K (Mezger, 1971; Heiles, 1971). Eq. (5) tells us that under these conditions gas clouds in the order of one solar mass are gravitationally unstable and con-

---

[a] If not otherwise specified in this paper extinction is expressed in magnitude and refers to the visual band at $\lambda = 5500$ Å.

tract. At least in some cases, fragmentation processes will lead to the simultaneous formation of stars of lower mass, as witnessed by the presence of many T-Tauri stars (believed to be protostellar objects) whose positions in the Taurus cloud are indicated by crosses in Fig. 12.

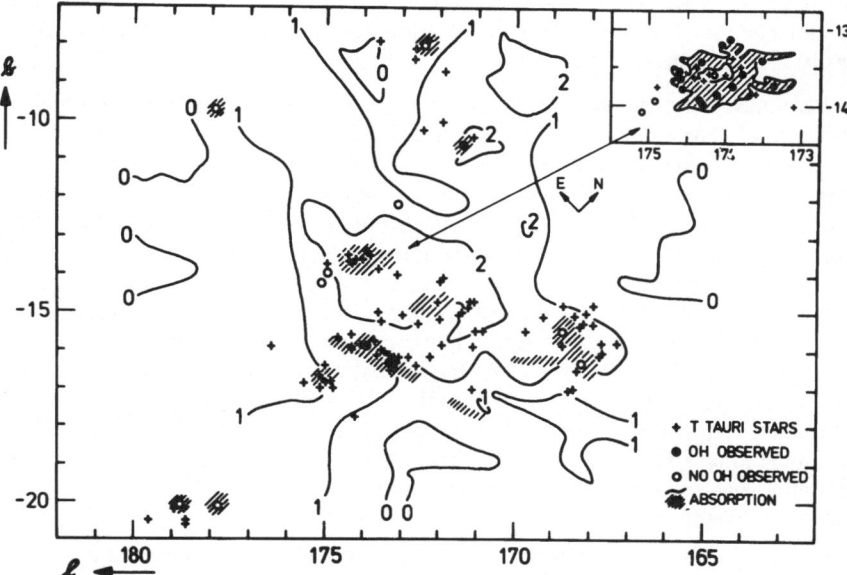

Fig. 12. Contour map of the visual extinction in the Taurus cloud. The shaded areas represent "dark clouds" of small diameter with visual extinction ranging from $5^m$ to $8^m$. Also indicated are the positions of T-Tauri stars and quasi-thermal OH emission

Dark clouds have been catalogued most recently by Lynds (1962). Although their distances are rarely known, Hollenbach et. al. (1971), with some plausible assumptions, have estimated physical parameters for Lynds dust clouds. Their results are given in Table 4.

Table 4. Parameters for the clouds in Lynds' catalogue (according to Hollenbach et al., 1971)

| $j$ | $N_j$ | $\Omega_j$ | $\dfrac{<R>}{pc}$ | $\dfrac{n_j}{cm^{-3}}$ | $\dfrac{<M>}{M_\odot}$ |
|---|---|---|---|---|---|
| 1...... | 215 | 0.125 | 6.65 | 48 | 1000 |
| 2...... | 312 | 0.17 | 6.5 | 100 | 1900 |
| 3...... | 520 | 0.175 | 5.0 | 200 | 1700 |
| 4...... | 560 | 0.061 | 3.0 | 440 | 800 |
| 5...... | 495 | 0.008 | 1.1 | 1500 | 135 |
| 6...... | 175 | 0.0017 | 0.85 | 2400 | 102 |

21

The data for the clouds are divided into categories $j = 1$ through $j = 6$, where $j$ is equal to the average extinction in magnitudes. If the clouds are assumed to be spherical, the optical depth to the cloud center $\tau_v$ is approximately $0.7 j$. $N_j$ is the number of $j$-type clouds observed, multiplied by 1.3 to allow for areas not covered by Lynds' survey, and $\Omega_j$ is the total solid angle subtended by the $N_j$ clouds in the $j$-th group. These clouds are estimated to lie between 100 and 1000 pc from the sun. The last three columns in Table 4 give the average radius, gas (number) density and total mass of the clouds.

Low-mass stars appear to be formed predominantly in dark clouds like the Taurus cloud. The total stellar mass spectrum ranges from stars of $0.08\,M_\odot$ to stars of mass about $50\,M_\odot$. In stars of mass $< 0.08\,M_\odot$, central temperatures never attain values at which hydrogen-burning will be initiated. What determines the upper limit for stellar mass of $\simeq 50\,M_\odot$ is not yet fully understood. The mass spectrum of stars decreases rapidly with increasing stellar mass. This indicates that it becomes increasingly difficult for massive stars to form out of the interstellar matter. O-stars of mass $M \geqslant 15\,M_\odot$ have effective (surface) temperatures $T_{eff} > 30,000\,°K$ and therefore emit a large fraction of their total radiation in the far UV at wavelengths $\lambda < 912\,Å$, the ionization limit for hydrogen. These massive stars, therefore, are capable of ionizing a large fraction of the surrounding interstellar gas, and thus form HII regions which can be observed in the radio and optical ranges. HII regions are at present the best indicators for regions in interstellar space where massive stars (the so-called OB-star clusters and associations) are formed.

In dark clouds some O-stars occasionally form and, in turn, generate a small HII region like the Orion Nebula (Fig. 1). Formation of low-mass stars over an extended period of time $> 10^7$ yr and the occasional appearance of an O-star seems to be typical for star formation in material arms. In genuine density-wave spiral arms in which the inflowing interstellar gas is compressed by as much as a factor of eight, the formation of O-stars occurs at a much higher rate and correspondingly the "giant HII regions" observed in density-wave spiral arms require a larger number of O-stars to account for their ionization. Which are the types of clouds out of which giant HII regions and their associated O-stars form? This question has been investigated by Scoville (1972) by means of the CO $\lambda$ 2.6 mm line. He finds that the column density of hydrogen inferred from observations of the CO-line ranges from $10^{22}$ to $3 \times 10^{23}$ H atoms cm$^{-3}$. Such clouds would have a visual extinction between $6^m$ and $180^m$ and the term "black clouds" used by Scoville to discriminate these objects from the "dark clouds" in regions like the Taurus cloud is certainly justified. The linewidths observed in black clouds are 4 to 40 times wider than the linewidths of about 1 km/sec typical of dark clouds. These large linewidths appear to be the result of large-scale motions.

Observations made with high angular resolution revealed for one of the black clouds in the nuclear disk a quasi-rigid-body rotation. Kinetic gas temperatures in black clouds appear to be higher than in dark clouds, which implies that they contain some source of heating. The average density in black clouds is of the order of $10^3$ particles cm$^{-3}$ but, in some condensations, may go as high as $10^5$

to $10^7$ particles $cm^{-3}$. Radii of black clouds range from 10 to 30 pc and masses range from $10^4$ to $10^6$ $M_\odot$.

## E. Summary

Interstellar matter accounts for about 10% of the total mass of the Galaxy. It forms a flat layer about the galactic plane and takes part in the general rotation of the stars around the center of the Galaxy. The average density of the interstellar gas in the galactic plane is between 0.3 and 0.7 atoms $cm^{-3}$. Dust particles of typical size $0.02\,\mu$ to $0.15\,\mu$ but of unknown composition are well mixed with the interstellar gas. H and $^4$He account for 98% of the total mass of the gas. It appears that the chemical composition of the interstellar gas as given in Table 2 is rather uniform throughout the Galaxy (and possibly throughout the universe).

Both density and kinetic temperature of the interstellar gas vary over a wide range. It appears to be a basic characteristic of the interstellar gas that dense ($n_H \geqslant 10$ $cm^{-3}$) and cool ($T_k < 100\,°K$) cloudlets of some $10\,M_\theta$ are embedded in a hot ($T_k > 1000\,°K$) and tenuous ($n_H \leqslant 0.2$ $cm^{-3}$) gas. Apart from this cloudlet structure, there exists a large-scale structure in the interstellar gas including material and densitywave spiral arms and the nuclear disk. Within these features, the average gas density is higher by a factor of about ten than that of the interarm gas. It is predominantly in these large scale features that other types of clouds, *i.e.* ”dark“ and ”black“ clouds, are observed. These clouds are much more massive an and have much higher densities than the cloudlets. Their kinetic gas temperature is usually very low. These clouds are the regions where stars are born and also where most of the interstellar molecules are observed.

Interstellar space is filled with radiation, both high energy particles (cosmic rays) and photons whose spectrum ranges from γ-rays and X-rays through the UV to the microwave region. In the context of interstellar molecules two components of this latter radiation are of special importance: the UV radiation with wavelengths longer than 912 Å which controls both formation and destruction of interstellar molecules, and the 2.7 °K black-body radiation, a relic of the "big bang" which in the absence of other interactions such as collisions determines the Local Thermodynamical Equilibrium (LTE) of molecules. Soft X-rays and subcosmic particles may contribute to the heating of the interstellar matter.

A more detailed compilation of observations pertaining to the interstellar gas is given by Mezger (1972).

## III. Observations of Interstellar Molecules

### A. Introduction

Within the last four years, nearly 30 different molecules have been identified in the interstellar gas clouds of our Galaxy. This advance has been made possible in part by improved techniques of radio astronomy, and has added a large variety of new interstellar molecules to the list of the three radicals CN, CH, and CH$^+$, known from their ultraviolet spectra since before 1940 (Adams, 1947; Swings and Rosenfeld, 1937; McKellar, 1940). The discovery in 1963 of the λ18 cm radio spectrum of OH by Weinreb *et al.* was the first identification of an interstellar molecule by radio astronomy. Thus, up until 1968, only four interstellar molecules were known to exist, leading to the generally accepted conclusion that simple free radicals were the main interstellar molecular constituents in a highly dilute gas ($<$ 1 particle cm$^{-3}$), subject to ionizing ultraviolet radiation.

However, the detection of radio spectral lines in the frequency range of about 22 to 23 GHz from the polyatomic molecules $NH_3$ and $H_2O$ (in 1968 and 1969) by Cheung *et al.* and the discovery of the organic molecule $H_2CO$ (in 1969) by Snyder *et al.* marked the beginning of a long series of discoveries. From these and all subsequent discoveries it became evident that dense and cool condensations of the interstellar matter are particularly rich in molecules. It is in the "dark" and "black" clouds (described in Section II. D) that rather complex organic molecules are being found. Fig. 13 presents the electromagnetic spectrum from about 1 Å ($10^8$ cm$^{-1}$) to $10^{10}$ Å ($10^{-2}$ cm$^{-1}$). Shown schematically are atmospheric absorption, the different techniques employed by astronomy and the various molecular effects occurring throughout this range. For wavelengths shorter than the 3000 to 10 000 Å "visible window", atmospheric absorption in the upper layers (higher than 50 km) of the atmosphere is very strong and effectively blocks all transmission. With the exception of a few well-defined windows between $1\mu$ and $24\mu$, the infrared radiation is mainly absorbed by water and carbon dioxide in the lower layers of the atmosphere below 30 km. In the radio region, rotational transitions of $H_2O$ and the pressure-broadened spin-reorientation transitions of molecular oxygen, $O_2$, cause atmospheric absorption in some well-defined bands. With the exception of these regions, the transmission is good beginning at about 1 mm and reaching out to a few meters where the ionosphere becomes opaque. The development of balloon and airplane techniques is now opening up the infrared region to astronomical observations, while rockets and satellites are making the ultraviolet region of the electromagnetic spectrum accessible for the first time. Furthermore, as is indicated in Fig. 13, different types of molecular transitions occur in each region of the electromagnetic spectrum: electronic transitions in the ultraviolet region, vibrational transitions in the infrared, and pure rotational transitions throughout the far infrared, the millimeter- and microwave regions.

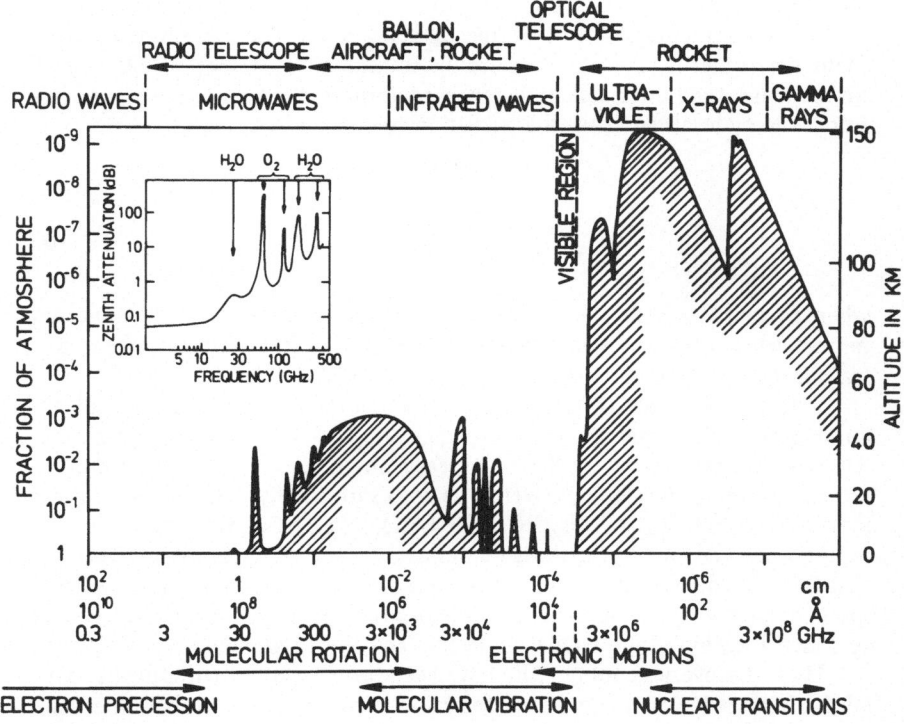

Fig. 13. Attenuation of electromagnetic radiation in the atmosphere. The solid curve indicates the altitude where the intensity of the external radiation is reduced to 1/2 of its original value at a given wavelength (after Giacconi *et al.*, 1968). Physical phenomena and observational techniques are indicated for the different wavelength regions. All observed interstellar radio frequency lines lie within the atmospheric windows

A molecular transition arises from a particular pair of energy levels. The energies of these levels depend upon the electronic configuration, the geometry, the masses, and the force field of the molecule. In the radio region, the majority of the transitions occur between rotational energy levels which correspond, classically speaking, to the end-over-end rotation of the molecules. For radio astronomical observation only rotational levels within the electronic ground state, which for most stable molecules is a $^1\Sigma$ state, are sufficiently populated to be detected. To a first approximation the rotational levels have a characteristic grouping which is determined by the molecular geometry, that is whether the molecule is a linear, a symmetric-top, or an asymmetric-top molecule. In Table 6 (to be discussed later) the molecules OCS, HCN, and HCC—CN are linear molecules for which, as for diatomic molecules, the principal moment of inertia about the bond axis is zero. $NH_3$, $CH_3CN$, and $CH_3CCH$ are symmetric-top molecules in which two of the three principal moments of inertia

25

are equal. All the other molecules are asymmetric-top molecules, with none of the three principal moments of inertia being equal. The magnitude of the moments of inertia (their inverses are the rotational constants), which depend on the masses and internuclear distances, determine the location of the rotational energy levels and thus the transition frequencies

$$\nu = \frac{E_u - E_l}{h} \tag{6}$$

where $E_u$ and $E_l$ are the energies of the upper and lower states of the molecule, respectively. Fortunately, however, not every pair of levels can give rise to a transition; only certain pairs do so, as governed by the selection rules.

With the exception of the OH and $NH_3$ transitions, all interstellar molecular lines observed to date in the radio range are pure rotational transitions. Although the microwave spectra of a great many molecules were studied in the laboratory during the last twenty years, a surprisingly large number of molecules including $H_2CS$, $H_2CNH$, $NH_2CHO$, NCO, HCO, HNO have only recently been detected or investigated in any detail in the laboratory. The first three of these molecules have subsequently been detected in interstellar space, emphasizing the importance of laboratory spectroscopy for the unequivocal identification of interstellar molecules and the assignment of their spectra.

These discoveries could hardly have been brought about without the extensive knowledge of energy levels, selection rules, transition probabilities and spectra assembled by laboratory spectroscopy for a very wide range of molecules. Microwave spectroscopy in the centimeter- and millimeter-wave regions is a well-established research field; for further information the reader is referred to the books by Gordy and Cook (1970), and Townes and Schawlow (1955), and to a recent review on millimeter-wave spectroscopy by Winnewisser *et al.* (1972). Reviews on interstellar molecules are given by Snyder (1972) and Rank *et al.* (1971).

## B. Observations

### 1. Theory of Line Emission and Absorption

Consider a gas which is exposed to radiation at the resonant frequency $\nu$. The molecules undergo transitions between the lower and the upper energy levels. In the case of statistical equilibium the number of molecules per unit time undergoing transitions from the lower state to the upper state (= absorption) equals the number of molecules making the reverse transition (caused by spontaneous and stimulated emission)

$$n_l \cdot u_\nu \cdot B_{lu} = n_u \left( u_\nu B_{ul} + A_{ul} \right) \tag{7}$$

where $n_u$ and $n_l$ represent the number of molecules in the upper and lower energy levels, respectively, and $u_\nu$ is the energy density of the radiation field. $B_{lu}$ represents the Einstein coefficient of absorption, whereas $B_{ul}$ and $A_{ul}$ are the Einstein coefficients of induced and spontaneous emission. Quantum mechanical calculations show that for isotropic radiation and non-degenerate energy levels

$$B_{lu} = \frac{8\pi^3}{3h^2} \ |\mu_{lu}|^2 = B_{ul} \tag{8}$$

where $\mu_{lu}$ is the dipole moment matrix element. Introducing the degeneracy of the levels with $g_u$ and $g_l$ as the statistical weights of the levels u and l it follows from Eq. (8)

$$g_l \cdot B_{lu} = g_u \cdot B_{ul} \tag{9}$$

Furthermore the Einstein coefficient $A_{ul}$ may be given in terms of the $B$ coefficients by use of Planck's radiation law

$$A_{ul} = \frac{8\pi h \nu^3}{c^3} \cdot \frac{g_l}{g_u} \cdot B_{lu} \tag{10}$$

and with Eqs. (8) and (9) one obtains

$$A_{ul} = \frac{64\pi^4 \nu^3 |\mu_{ul}|^2}{3h c^3} \tag{11}$$

The absorption coefficient $\alpha'_\nu$ the fractional change in energy density per unit length, is $\frac{h\nu}{c}$ times the net absorption per cm³ per second[b]

$$\int_{\nu=0}^{\infty} \alpha'_\nu \, d\nu = \frac{h\nu}{c} \left\{ n_l B_{lu} - n_u B_{ul} \right\} \tag{12}$$

$$= \frac{h\nu}{c} n_l B_{lu} \left\{ 1 - \frac{n_u}{g_u} \cdot \frac{g_l}{n_l} \right\}$$

In the case of local thermodynamical equilibrium (LTE) at a temperature $T$, the distribution of the molecules over different states is given by the Boltzmann distribution:

$$\frac{n_u}{g_u} \cdot \frac{g_l}{n_l} = e^{-\frac{h\nu}{kT}} \tag{13}$$

---

[b] In astrophysics this quantity is often referred to as "absorption coefficient, corrected for stimulated emission".

However, if there is no thermodynamic equilibrium, as is common for interstellar cases, it is convenient to define between two energy levels an excitation temperature $T_{ex}$ which is determined by fitting the observed molecular distribution to the Boltzmann formula, Eq. (13).

Thus, the correction term for stimulated emission becomes

$$1 - \frac{n_u}{g_u} \cdot \frac{g_l}{n_l} = 1 - e^{-h\nu/kT_{ex}} \tag{14a}$$

$$\simeq 1 - (1 - \frac{h\nu}{kT_{ex}} + \ldots) \simeq \frac{h\nu}{kT_{ex}}$$

$$\text{for } h\nu \ll kT_{ex} \tag{14b}$$

Or in convenient units for numerical calculations, where $T_{ex}$ is measured in units of the absolute temperature scale (°K)

$$T_{ex} \gg \frac{h\nu}{k} = 4.8 \times 10^{-2} \cdot \left[ \frac{\nu}{GHz} \right] = 1.44 \left[ \frac{1/\lambda}{cm^{-1}} \right] \tag{15}$$

Excitation temperatures in interstellar molecular clouds are typically $3\,°K \leqslant T_{ex} \leqslant 50\,°K$, and thus the approximation (14b) looses its validity in the shorter mm-wavelength region.

The observable quantity in astronomical work, however, is not the absorption coefficient per unit length, but the opacity or optical depth $\tau_\nu$, i.e. the absorption coefficient $\alpha'$ integrated over both the entire path length $L$ along the line of sight and the line profile

$$\int_0^\infty \tau_\nu \, d\nu = \int_0^L dl \int_0^\infty \alpha'_\nu \, d\nu = \frac{c^2}{8\pi\nu^2} \cdot \frac{g_u}{g_l} \left( 1 - e^{-\frac{h\nu}{kT_{ex}}} \right) \cdot N_l \tag{16}$$

This equation is obtained by using equations (III. 6, 7 and 8). Here $N_l = \int_0^L n_l dl$ represents the number of molecules in the lower energy level contained in a column of cross section 1 cm$^2$ along the line of sight. This quantity is referred to as the column density of the molecules in the lower level and is directly obtained from astronomical observations.

$N_l$ can, however, be related to the total column density, $N = \int_0^L n dl$, i.e. the total number of molecules per cm$^{-2}$ along the line of sight, through the rotational partition function $Q_r$, by the following expression

$$N_l = N \cdot g_l \cdot \frac{\exp(-E_l/kT_{ex})}{Q_r} \tag{17}$$

where $E_1$ is the energy of the lower level 1 and

$$Q_r = \sum_i g_i e^{-E_i/kT_{ex}} \tag{18}$$

where the summation extends over all quantum numbers. The latter quantity is calculated under the assumption that the population of all rotational levels is determined by a Boltzmann distribution with $T = T_{ex}$, $i.$ $e.$ one assumes local thermodynamic equilibrium (LTE). In interstellar space, however, observational results show more or less strong deviations from LTE.

Evaluation of the partition function in the general case of an asymmetric rotor molecule is rather tedious. However, for rigid diatomic or linear molecules characterized by the rotational constant $B$ (in MHz), the partition function can be calculated by replacing the summation over the total quantum number $J$ by an integral

$$Q_r \approx \int_0^\infty (2J+1) \exp\left(-\frac{hBJ(J+1)}{kT_{ex}}\right) dJ \simeq \frac{kT_{ex}}{hB} \quad \text{for } kT \gg hB \tag{19}$$

which is a good approximation even for interstellar conditions. Now a general relation between the integrated optical depth of a molecular line and the total column density $N$ can be derived by substituting Eqs. (17) into Eq. (16)

$$\int_0^\infty \tau_\nu \, d\nu = \frac{c^2}{8\pi\nu^2} A_{lu}\left(1 - e^{-h\nu/kT_{ex}}\right) g_u \frac{\exp(-E_1/kT_{ex})}{Q_r} \cdot N \tag{20}$$

Thus, the determination of the total column density $N$ depends heavily on an accurate determination of $T_{ex}$, the optical depth $\tau_\nu$ and the evaluation of the partition function $Q_r$.

The determination of $T_{ex}$ and its interpretation in terms of the physical conditions which exist in interstellar clouds is an intriguing task of molecular radio astronomy. Two limiting cases are readily considered: i) if the rate of collisionally induced transitions $C_{ul}$ is small compared to the radiative rate $A_{ul}$ then $T_{ex}$ is determined by the 2.7 °K radiation field and ii) if the reverse is true, then $T_{ex} \approx T_k$, the kinetic temperature of the gas

$$T_{ex} = \begin{cases} T_{bb} = 2.7\,°K & \text{for } C_{ul} \ll A_{ul} \\ \\ T_k & \text{for } C_{ul} \gg A_{ul} \end{cases} \tag{21}$$

Typical kinetic gas temperatures in dense interstellar clouds are $T_k \leqslant 30\,°K$ but may in some cases reach values of $\sim 80\,°K$ (e. g. the Orion molecular cloud). The relation between excitation temperature $T_{ex}$ and kinetic gas temperature $T_k$ is further discussed in Sections III. E and III. F.

## 2. Molecular Lines in the Optical and UV Range

The electronic spectra of molecules are always seen in absorption against the continuum spectrum of a bright background star with high effective temperature. Strong extinction of the starlight by intervening dust clouds often makes the background star too faint for observation; this excludes from optical observations the molecule rich dark clouds commonly observed by radio techniques. With the notable exception of CN and $H_2$, all detected molecular absorption lines in the optical or UV region are due to transitions from the lowest rotational level of the vibrational and electronic ground state to various (rotationally and) vibrationally excited energy levels within a higher electronic state. In this case the correction factor for stimulated emission becomes unity, and Eq. (20) reduces to

$$\int_0^\infty \tau_\nu \, d\nu = \frac{c^2}{8\pi\nu^2} A_{lu} \frac{g_u}{g_l + g_u} \cdot N_{tot} \tag{22}$$

In the optical region a detailed measurement of the line profiles is usually difficult, and for absorption lines it has therefore become usual to measure an absorption line by its equivalent width $W_\nu$ defined in frequency units ($W_\nu = \frac{\nu}{\lambda} \times W_\lambda$)

$$W_\nu = \int (1 - \frac{I_\nu}{I_0}) \, d\nu = \int (1 - e^{-\tau_\nu}) \, d\nu \tag{23a}$$

where $I_\nu$ is the intensity in the line and $I_0$ is the intensity in the adjacent continuum spectrum. The general relation between $W_\nu$ and $N_l$ is called the "curve of growth" which, depending on the value of the optical depth, has three characteristic parts. The equivalent width is i) on the "linear section" if $\tau_\nu \ll 1$, in which case Eq. (23a) can be expanded to yield with Eq. (22) a linear relationship between $W_\nu$ and $N$

$$W_\nu = \int \tau_\nu \, d\nu \propto N \tag{23b}$$

and ii) on the "flat section" if $\tau_\nu > 1$. Then the exact formula Eq. (23a) has to be used and the functional dependence of the curve of growth is determined by the detailed shape of the profile function iii). If the lines become very strong ($\tau_\nu \gg 1$) the line profile function should be determined by the radiation damping wings, and the line is in the "square root section" of the curve of growth for which

$$W_\nu \propto \sqrt{N} \tag{23c}$$

The Lyman lines of HI and the Lyman resonance bands of $H_2$ fall in the latter category. The column density of $H_2$ is then obtained (Carruthers, 1970) by comparing the equivalent widths measured in the stellar spectrum and in the laboratory spectrum, from which a column density $N_{H_2} = 1.3 \ (+0.9, -0.65) \times$

Table 5. Molecules detected by optical techniques

| Molecule | Transition | Wavelength Å | Other transitions | Column densities | Source | Year of discovery |
|---|---|---|---|---|---|---|
| $^{12}$CH | $A^2\Delta - X^2\Pi$ (0,0) $R_2$ (1) | 4300 | Y | $\sim 10^{13}$ | $\sim 40$ | 1937 |
| | $B^2\Sigma^- - X^2\Pi$ (0,0) $^PQ_{12}$ (1) | 3890 | Y | | | 1941 |
| | $C^2\Sigma^+ - X^2\Pi$ (0,0) $^PQ_{12}$ (1) | 3146 | Y | | | 1960 |
| $^{12}$CH$^+$ | $A^1\Pi - X^1\Sigma$ (0,0) R (0) | 4233 | Y | $10^{13}$ | $\sim 60$ | 1937 |
| $^{13}$CH$^+$ | $A^1\Pi - X^1\Sigma$ (0,0) R (0) | 4232 | Y | $10^{13}$ | $\sim 2$ | 1969 |
| CN | $B^2\Sigma^+ - X^2\Sigma^+$ (0,0) R (0) | 3875 | Y | $10^{12}$ | 14 | 1938/39 |
| CO | $A^1\Pi - X^1\Sigma$ (1,0) | 1510 | Y | $10^{15}$ | $\zeta$ OPH | 1971 |
| | $C^1\Sigma^+ - X^1\Sigma^+$ (0,0) R (0) | 1088 | Y | $\sim 10^{13}$ | 3 | 1973 |
| | $E'_\pi - X^1\Sigma^+$ (0,0) R (0) | 1076 | Y | $\sim 10^{13}$ | 3 | 1973 |
| $^{13}$CO | $A^1_\pi - X^1\Sigma$ (2,0) | 1476 | Y | $\sim 10^{13}$ | $\zeta$ OPH | 1971 |
| H$_2$ | $B^1\Sigma_u - X^1\Sigma_g$ (0,0) | 1108 | Y | $10^{20}$ | 15 | 1970 |
| HD | $B^1\Sigma_u^+ - X^1\Sigma_g^+$ (3,0) R (0) | 1066 | Y | $\sim 10^{14}$ | $\sim 9$ | 1973 |

$10^{20}$ cm$^{-2}$ has been derived in the direction of $\xi$ Persei. The large error margins are essentially determined by the low intensity of the continuum in the stellar spectrum which causes large uncertainties in the measurements of equivalent width.

Just as hydrogen is the most abundant element, so molecular hydrogen $H_2$ appears to be the most abundant molecule in space. In 1970 Carruthers dis-

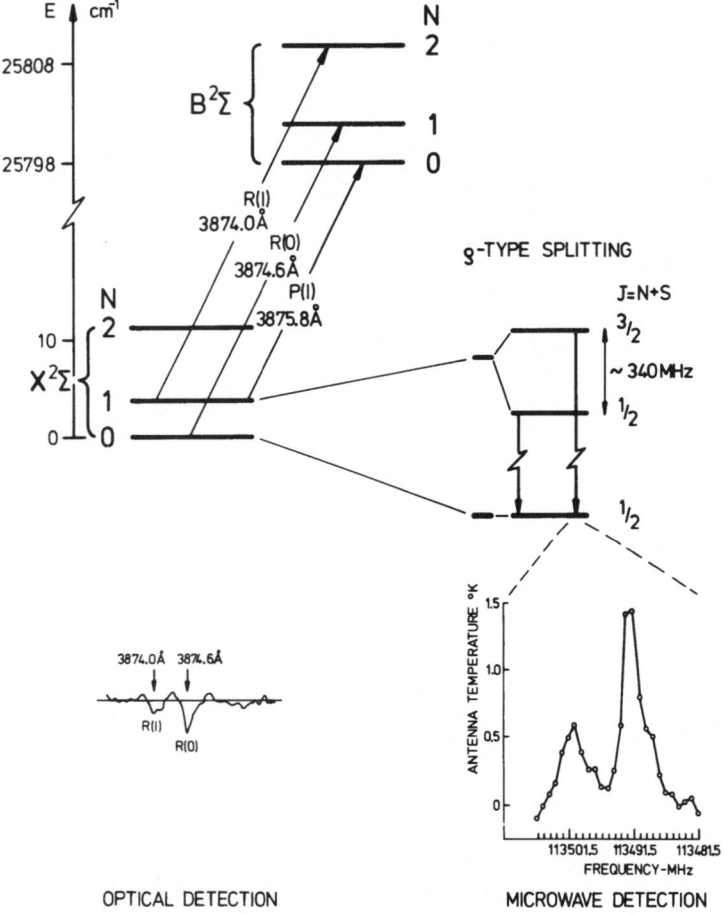

Fig. 14. Energy levels of the free radical CN showing the observed optical and radio transition. The violet band is observed against $\xi$ Ophiuchi (Thaddeus and Clauser, 1966) and the radio frequency transitions are measured in the Orion nebula (Jefferts *et al.*, 1970). In the interstellar optical spectra quadrupole hyperfine and spin-doubling structure are entirely unresolved and therefore the rotational levels can be approximated as though the electronic states were $^1\Sigma$. However, these splittings are partly resolved in the radio frequency region but we note that only one component of the spin doublet has been observed

covered the Lyman resonance absorption bands $B^1 \Sigma_u - X^1 \Sigma_g$ ($v'-v'' = 0-0$, . . . . . , 7–0) of interstellar $H_2$ (see Table 5) in the far ultraviolet spectrum of only one star, $\xi$ Persei, by employing special observing techniques such as windowless photo-multipliers and LiF coatings on all optics. Results from the OAO–Copernicus satellite confirm the suggested high abundance of the $H_2$ molecule (Spitzer *et al.* 1973).

Strong $H_2$ lines have been reported in 11 reddened stars ($E (B - V) > 0.10$). Considerably less or no $H_2$ absorption was detected in unreddened stars. Large column densities have been reported in higher rotational levels (up to $J = 6$) which correspond to an excitation temperature between 150° and 200 °K. The ratio of ortho-hydrogen ($J$ = odd) to para-hydrogen ($J$ = even) correspond to about 80 °K. Absorption of two lines of the Lyman bands of HD were also reported in nine stars, indicating a ratio of $HD/H_2 \sim 10^{-6}$. However, in the star $\beta$ Cen a deuterium to hydrogen ratio of $N(D)/N(H) = 1.047$ (0.125, −0.112) x $10^{-5}$ has been determined. (Rogerson *et al.* 1973).

CO has been detected in the interstellar absorption spectrum of $\zeta$ Ophiuchi and has thus become the second interstellar molecule to be detected by rocket ultraviolet spectroscopy. Smith and Stecher (1971) have detected eight transitions in the fourth positive system of $^{12}C^{16}O$ and four of $^{13}C^{16}O$, which yielded a $^{12}C/^{13}C$ ratio of 105. It seems likely that interstellar molecular detections in the vacuum ultraviolet will follow, especially of polyatomic molecules like $H_2O$.

The optical spectrum of CN provided the first determination for the measurement of the microwave background radiation temperature. The electronic transitions are not only seen from the lowest ground-state level ($N = 0$) but also from the first excited rotational state ($N = 1$). From the intensity ratio of the two lines R(0)/R(1) $\sim$ 5 McKellar (1940) derived a rotational temperature of 2.3 °K for the interstellar medium. This temperature is close to the measurement of 2.7 °K for the isotropic background radiation (Penzias and Wilson, 1965). Thus the significance of the optical transitions of CN is that they provide a measurement of the 2.7 °K background radiation at a wavelength of 2.6 mm. From a recent analysis of the R(1)/R(0) and P(1)/R(0) line ratios the presently most accurate value of the background intensity has been derived to be $T_{bb}$ (2.64 mm) = 2.78 ± 0.10 °K (Thaddeus, 1972). Fig. 14 summarizes both optical and radio results on CN, although it must be born in mind that these observations are done in different interstellar environments, *i.e.* in low and high-opacity clouds, respectively. From the energy level scheme it is clear that radio frequency detection of the lower spin component would confirm the assignment and simultaneously yield accurate ground state constants, not yet obtained from microwave measurements.

*Note Added in Proof.* The lower spin component $J = 1/2-1/2$ of the $N = 1 \rightarrow 0$ transition of CN has now been detected by Penzias *et al.*, 1973.

## 3. Radio Observations

In dark clouds the dust protects the molecules from dissociation by UV radiation. This protection is presumably even better in the case of Scoville's black

33

Table 6. Molecules detected by radio techniques

| Molecule | Isotopes | Transition | Energy cm$^{-1}$ | Frequency MHz | Other transitions | HFS | Emission/absorption | Column density cm$^{-2}$ | Number of sources | Year of discovery |
|---|---|---|---|---|---|---|---|---|---|---|
| OH hydroxyl radical | $^{18}$OH | $^2\Pi_{3/2}$; $J = 3/2$ | 2 | 1665 | Y | Y | E, A, ME | $10^{12} - 10^{16}$ | $> 200$ | 1963 |
| | | $^2\Pi_{1/2}$; $J = 1/2$ | 135 | 4766 | Y | Y | E | | 1 | 1968 |
| SiO silicon monoxide | | $J = 3 - 2$ | 9 | 130268 | Y | None | E | $10^{13}$ | 1 | 1971 |
| SO sulfur monoxide | $^3\Sigma^-$; | $J, N = 3,2 - 2,1$ | 8 | 99300 | | | E | $\sim 10^{15}$ | 7 | 1973 |
| H$_2$O water | | $J_{K_a K_c} = 6_{16} - 5_{23}$ | 447 | 22235 | | | ME | ? | $> 50$ | 1969 |
| H$_2$S hydrogen sulfide | | $J_{K_a K_c} = 1_{11} - 1_{10}$ | 19 | 168763 | | | E | $\sim 3 \times 10^{14}$ | 7 | 1972 |
| NH$_3$ ammonia | | $J, K = 2,2$ | 45 | 23723 | Y | | E | $\sim 10^{16}$ | $\sim 10$ | 1968 |
| CN cyanogen radical | | $^2\Sigma$; $N = 1 - 0$ | 3 | 113492 | | Y | E | $10^{15}$ | 3 | 1970 |
| CO carbon monoxide | $^{13}$C$^{16}$O $^{12}$C$^{18}$O $^{12}$C$^{17}$O | $J = 1 - 0$ | 4 | 115271 | | None | E | $10^{17} - 10^{19}$ | $> 50$ | 1970 |
| CS carbon monosulfide | $^{12}$C$^{34}$S $^{13}$C$^{32}$S | $J = 3 - 2$ | 10 | 146969 | Y | None | E | $10^{13} - 10^{14}$ | 4 | 1971 |
| OCS carbonylsulfide | | $J = 9 - 8$ | 18 | 109463 | Y | None | E | $10^{16}$ | 1 | 1971 |
| HCN hydrogen cyanide | H$^{13}$C$^{14}$N H$^{12}$C$^{15}$N D$^{12}$C$^{14}$N | $J = 1 - 0$ | 3 | 88632 | Y | Y | E | $10^{14} - 10^{15}$ | $\sim 10$ | 1970 |

| | | | | | | | | | |
|---|---|---|---|---|---|---|---|---|---|
| H₂CO formaldehyde (H₂¹³C¹⁶O, H₂¹²C¹⁸O) | $J_{K_aK_c} = 1_{10} - 1_{11}$ | 4830 | 11 | Y | Y? | A | $10^{12} - 10^{16}$ | >100 | 1969 |
| H₂CS thioformaldehyde | $J_{K_aK_c} = 2_{11} - 2_{12}$ | 3139 | 13 | | | A | $>10^{16}$ | 1 | 1971 |
| HNCO isocyanic acid | $J_{K_aK_c} = 4_{04} - 3_{03}$ | 87925 | 15 | Y | Y | E | $\sim 10^{14}$ | 2 | 1971 |
| HCOOH formic acid | $J_{K_aK_c} = 1_{10} - 1_{11}$ | 1639? | 3 | | | E | $<10^{13}$ | 1 | 1970 |
| HC₃N cyanoacetylene | $J = 1 - 0$ | 9098 | 0.2 | Y | Y | E | $10^{16}$ | ~2 | 1970 |
| H₂CNH methyleneimine | $J_{K_aK_c} = 1_{10} - 1_{11}$ | 5290 | 8 | Y | Y | E | $>10^{14}$ | ~1 | 1972 |
| CH₃OH methylalcohol | $J_{K_aK_c} = 1_{10} - 1_{11}$ | 834 | 12 | Y | | E, ME? | | ~3 | 1970 |
| CH₃CN methylcyanide | $J, K = 6,0 - 5,0$ | 110383 | 13 | Y | Y | E | $\sim 10^{14}$ | 2 | 1971 |
| HCONH₂ formamide | $J_{K_aK_c} = 2_{11} - 2_{12}$ | 4619 | 4 | | Y | E | $>10^{11}$ | ~2 | 1971 |
| CH₃CHO acetaldehyde | $J_{K_aK_c} = 1_{11} - 1_{10}$ | 1065 | 2 | | | E | $\sim 10^{14}$ | 1 | 1971 |
| CH₃CCH methylacetylene | $J, K = 5,0 - 4,0$ | 85457 | 3 | | | E | ? | 2 | 1971 |
| (X-ogen) | U(89190) | | | | | E | | >8 | 1971 |
| (HNC?) | U(90665) | | | | | E | | 5 | 1971 |

U stands for unidentified line

35

clouds (see Section II. D). However, unlike optical and UV radiation, radio waves are not affected by interstellar dust and can penetrate these clouds and interact with the molecules. The radio observations of molecules are our main source of information about conditions inside these clouds. In Table 6 we list all molecules identified to date in interstellar space by radio techniques and give a summary of the results obtained. The first column of Table 6 gives the chemical formula and the name, the second indicates other isotopes which have been detected. The next three columns indicate the spectroscopic assignment of the transitions, their frequencies for the most abundant isotopic species, and the energy of the upper of the two levels involved in the transition. The transitions listed are representative but not complete. The next column indicates by a Y if other transitions have been observed. We have added in Appendix a complete listing of all observed interstellar molecular transitions, along with their distribution amongst the six most productive molecular sources. Some transitions show hyperfine structure; when this has been observed it is indicated by a Y in the next column. In those cases the theoretical centre frequency is quoted. The last three columns contain some astrophysical information: whether the lines appear in absorption (A), emission (E), or maser emission (ME); the total column density along the line of sight; and the approximate number of sources in which the molecules have been detected. In the following a brief discussion of the physical significance of these quantities will be presented.

Radiation intensities are referred to the radiation of a black body. In the radio frequency range it is convenient to express line intensities in equivalent line brightness temperatures, since the surface brightness $B_\nu$ of a black body

$$B_\nu = \frac{2k}{\lambda^2} T_b \quad \text{for } h\nu \ll kT_b \tag{24}$$

is directly related to its brightness temperature $T_b$. This is the Rayleigh-Jeans approximation of Planck's radiation law, and is valid throughout the cm-wavelength range and — with a few exceptions — in the millimeter wavelength range as well.

In radio astronomy multichannel or autocorrelation (Fourier) spectrometers are used which simultaneously cover the whole line profile. Consider a molecular cloud observed against a source of continuum radiation of a given brightness temperature. The continuum brightness temperature is the sum of the 2.7 °K isotropic background radiation $T_{bb}$ of a continuum source (such as an HII region or a supernova remnant) which may be in the line of sight and located behind the molecular cloud. A specific molecular transition with optical depth $\tau_\nu$ and excitation temperature $T_{ex}$ has an observable brightness temperature $T'_L$ [c)]

$$T'_L = \{T_{ex}\,(1 - e^{-\tau_\nu}) + (T_{bb} + T_c)\,e^{-\tau_\nu}\} \qquad (25a)$$

Here, the first term represents the line emission proper from the molecular cloud. The second term describes the attenuation of the continuum radiation due to absorption by the molecules.

With the center frequency of the spectrometer shifted outside the line profiles the measurement yields the unattenuated continuum temperature

$$T'_c = T_{bb} + T_c \qquad (25b)$$

The line profile proper is obtained by subtraction of (25b) from (25a)

$$T_L = T'_L - T'_c = \{T_{ex} - (T_{bb} + T_c)\}\,(1 - e^{-\tau_\nu}) \qquad (25c)$$

It is readily seen from Eq. (25c) that if $T_{bb} + T_c > T_{ex}$, the molecular line will appear in absorption, whereas if $T_{bb} + T_c < T_{ex}$, the line will appear in emission. If there is no continuum source in the line of sight, $T_c = 0$, and if collisional transitions are slow compared with radiative transitions induced by the isotropic background radiation, the molecules equilibrate with the 2.7 °K radiation, $T_{ex} = T_{bb}$, and the observable line brightness temperature according to Eq. (25c) is $T_L = 0$. Therefore, if a molecular line is seen in emission its excitation temperature $T_{ex} > T_{bb}$. This condition imposes a lower limit on the gas density of the emitting cloud (Section III. E). The continuum temperature of galactic radio sources decreases rapidly with frequency ($T_c \propto \nu^{-2.1}$ for thermal sources and typically $T_c \propto \nu^{-2.7}$ for nonthermal sources). Therefore, molecular lines are seen in absorption only in the cm-wavelength range, whereas all molecular lines in the mm-wavelength range are seen in emission (see Table 6).

If $\tau_\nu \gg 1$ it follows from Eq. (25c) that $T_L = T_{ex}$, i.e. in this case the observation yields directly the excitation temperature of the line. For optically thin clouds ($\tau_\nu \ll 1$) seen in emission one obtains

$$T_L \simeq (T_{ex} - T_c - T_{bb}) \cdot \tau_\nu \qquad (26)$$

Furthermore, one can generally use approximation (14b) for the correction factor for stimulated emission. Hence according to Eqs. (16 and 20) the optical depth of radio molecular lines is inversely proportional to the excitation temperature

---

[c)] The observable quantity in radioastronomy is the antenna temperature $T_{A,L}$ which is related to the brightness temperature $T'_L$ by $T_{A,L} = \eta_B T'_L$, provided the solid angle $\Omega_s$ subtended by the cloud is large in comparison to the antenna solid beam angle $\Omega_A$. Here $\eta_B$ is the beam efficiency of the radio telescope which is typically $\eta_B \simeq 0.7$. A more general relation is $T_{A,L} = \eta_B T'_L \Omega_s/\Omega_A$.

$$\int_0^\infty \tau_\nu \, d\nu \propto \frac{1}{T_{ex}} \int_0^\infty n_1 \, dl = \frac{N_1}{T_{ex}} \qquad (27)$$

Therefore, the fractional column density $N_1$ (and the total column density $N$ according to Eq. (20)) can only be obtained from the optical depth if $T_{ex}$ is known. If a gaussian line shape is assumed, integration of the observed line profile yields

$$\int_0^\infty T_L \, d\nu = T_{L,max} \times 1.065 \, \Delta\nu = (T_{ex} - T_c - T_{bb}) \times \int_0^\infty \tau_\nu \, d\nu \qquad (28)$$

Here, $T_{L,max}$ is the peak temperature of the line and $\Delta\nu$ the total line width between half intensity points. Substitution of Eq. (27) in (28) shows that

$\int_0^\infty T_L d\nu \propto N_1$ if $T_{ex} \gg |T_{bb} + T_c|$. Thus for optically thin lines in emission it

becomes possible to evaluate the fractional column density $N_1$ from the observed integrated line profile. However, in the case of rotational lines the transition from the fractional to the total column density according to Eqs. (19) and (20) again requires the knowledge of $T_{ex}$. Thus the general uncertainties in the determination of column densities are usually rather large. The column densities derived from emission lines are more reliable than those obtained from absorption lines. In case of strong deviations from LTE (such as the OH or $H_2O$ maser emission sources) no column densities can be estimated at all. All the uncertainties inherent in the determination of column densities put large error margins on the quantitative values given in column 9 of Table 6. These error margins are further increased if column densities are converted into volume densities using the relation $n = N/L$, owing to uncertainties in the estimated depth $L$ of the molecular cloud. However, some conclusions can be drawn. For example a comparison of the total column densities presented in Table 6 suggests indeed that free radicals like OH or CN are by no means overabundant with respect to stable diatomic molecules like CO or CS or even polyatomic molecules like $H_2CO$, HCN, or $CH_3CN$.

For four molecules various isotopically substituted species have been observed. Besides observations of $^{13}C$, $^{18}O$ and $^{17}O$ substitutions in CO and $H_2^{13}CO$, $H_2C^{18}O$ and $^{18}OH$.

Wilson et al. (1972) have recently reported the first positive $^{15}N$ isotopic detection in an astronomical source. They observed in the Orion molecular cloud the $J = 2 \to 1$ transitions of the rare hydrogen cyanide isotopic species $H^{13}C^{14}N$ and $H^{12}C^{15}N$ whose rest frequencies are 172 677.7 MHz and 172 108.1 MHz, respectively (Winnewisser et al., 1971). For these two molecules and in almost all the other isotopic abundance determinations, which can presently be given only with large error margins ($\sim 20\% - 50\%$), the isotopic ratios are found to have essentially terrestrial values.

There seem, however, to be two exceptions:

i) in the center region of our galaxy (Sgr A and Sgr B2) the $^{12}C/^{13}C$ ratios are most likely to be in the range from 20 to 50 (Whiteoak and Gardner, 1972), which is smaller than the terrestrial value of 89.

ii) deuterium has not been detected in interstellar space until very recently, when Jefferts $et~al.$ (1972) reported the detection of the two lowest rotational transitions of DCN in the Orion molecular cloud. They found the emission of the $J = 2 \rightarrow 1$ line of $D^{12}C^{14}N$ to be almost as intense as the $H^{12}C^{15}N$ line. Previously a $^{15}N/^{14}N$ ratio of $\sim 1/300$ was derived (Wilson $et.~al.$, 1971), thus implying that in the Orion molecular cloud the D/H ratio of hydrogen cyanide is about the same value, which is an order of magnitude greater than the terrestrial value. However, direct determinations of the interstellar D/H ratio by means of the $\lambda\,92$ cm hyperfine structure line of atomic deuterium (analog to the $\lambda\,21$ cm line of atomic hydrogen) by Weinreb (1962) and Cesarsky $et~al.$ (1972) are more in agreement with the terrestrial abundance. The disagreement between the ratios DCN/HCN and D/H probably indicates that some form of chemical fractionation is taking place.

## C. Identification of Interstellar Molecules

Because of the high precision with which the frequencies of the interstellar lines can be measured (better than 1 part in $10^5$) there remains usually little doubt about the positive identification of the molecular species, despite the fact that only a few transitions out of the whole rotational spectrum of any one given molecule have been observed to date in the radio frequency range. Confirmation is obtained from observations of other rotational transitions, or from the detection of possible fine-structure components, or from observations of corresponding transitions of isotopically substituted species. However, some uncertainty still remains in the identification of formic acid, HCOOH, whose $1_{10}-1_{11}$ transition is located in between two $^{18}OH$ resonances. An independent search for the $1_{01}-0_{00}$ transition for formic acid was negative (Snyder and Buhl, 1972). Similarly the identification of $H_2S$ and $H_2O$ still rests on only one observed interstellar radio transition and awaits further confirmation by the detection of other transitions.

With the exception of CN, all molecular identifications of interstellar lines are based on direct laboratory measurements. The location of the CN radio transitions can be predicted from laboratory ultraviolet data (Poletto and Rigutti, 1965) and is found to be in reasonable agreement with the positions given by the interstellar observations. However, the hyperfine structure (electric quadrupole and nuclear spin-electronic spin interaction) cannot be uniquely assigned; therefore an element of caution has to be maintained until more observational evidence — either from interstellar or laboratory measurements — confirms the assignment. (See Note Added in Proof p. 33).

There is also an increasing number of radio lines for which identification based on laboratory spectra has not been possible (see Table 6) suggesting that some of these lines may come from molecular transients, possibly radicals or even molecular ions difficult to obtain under laboratory conditions.

The positive identification of the carrier of these unassigned lines is of extreme importance to a better understanding of the formation mechanisms presently under discussion (Section IV).

However, by gathering more interstellar experimental data on other possible transitions, the assignment of these lines could be obtained by making use of the combination principle, which is often used to unequivocally assign complicated laboratory spectra. Furthermore, for one particular molecule and in one given cloud, the detection of as many molecular transitions as possible is required for rather different reasons. A detailed interpretation of the physical conditions (LTE and non-LTE effects) existing in a molecular cloud will only become possible if many molecular levels — and preferably connecting ones — have been studied by detection of the corresponding transitions between them. Molecular rotational transitions occur over a very wide wavelength region (see Fig. 13). It will therefore be necessary to cover not just a small range but rather the entire portion of the radio spectrum accessible to ground-based radiotelescopes. The millimeter-wave region will be particularly useful since

i) molecules which can be studied in the centimeter-wave region have usually not only more but also stronger lines at higher frequencies (see III. D) and

ii) light molecules such as CO, CN, HCN, $H_2S$, $H_2O_2$ and others have few if any transitions below 60 GHz.

These are the principal reasons why more than 50 % of all interstellar molecular lines detected to date lie in the millimeter-wave region, although the observing equipment available is much less sensitive for the millimeter-wave region than for the cm-wavelength range.

Furthermore it remains to be pointed out that many molecules of potential astrophysical interest have been sought in interstellar clouds but have not been found. Some of the notable negative results include cyclic molecules and NO, $H_2C_2O$ and others. However, at the present time it seems premature to draw definite conclusions since the detection limits are barely below the expected line intensities. We may note that $CH_4$ which is expected to exist in interstellar space (Section IV) has no allowed pure rotational spectrum. However, Dorney and Watson (1972) have shown that a centrifugally induced dipole moment exists which produces a complicated forbidden rotational spectrum in the radio frequency and microwave regions. The particular transitions are very weak and have not yet been observed in the laboratory.

## D. Laboratory Spectroscopy Relevant to Astronomical Observations

Observations of interstellar rotational transitions of $CH_3OH$ and $H_2S$ at 0.8 GHz and 169 GHz, respectively, indicate the wide frequency range now available to radio astronomy. Extension of earthbound observations to 300 GHz can be expected in the near future. The development of sensitive narrow-band IR detectors in conjunction with high-flying airplanes, balloons and the post-Apollo space shuttle program will probably open the way for spectroscopy in the far IR. All these observations require, however, extensive laboratory work, both for successful planning and for interpretation of the observational data. We discuss below some aspects of laboratory microwave spectroscopy pertaining to the interpretation of interstellar molecular radio transitions.

In microwave and millimeter wave laboratory spectroscopy an essentially monochromatic source is tuned over the spectral line to be measured. The output power of this source is collimated through an absorption cell containing the sample gas at low-pressure ($10^{-2}$ to $10^{-5}$ mm Hg) and is focussed onto the detector. The molecular absorption is indicated as a sharp decrease in the voltage output of the detector. The intensity of a microwave transition may be represented by the absorption coefficient $\alpha'(\nu)$ per unit length. An expression for this quantity is obtained by combining Eqs. (9 through 13) and assuming $h\nu \ll kT$, as is generally the case for the radio frequency region

$$\alpha'(\nu) = \frac{8\pi^3 n_1 \nu^2 |\mu|^2}{3ckT_{\text{ex}}} \cdot f(\nu) \qquad (29)$$

Here, $n_1$ is the number of molecules per $cm^3$ in the lower level, $\mu$ the electric dipole transition matrix element, $T_{\text{ex}}$ the excitation temperature of the molecule and $f(\nu)$ the line shape function normalized to

$$\int_0^\infty f(\nu)\, d\nu = 1.$$

The other quantities have their usual meaning.

If the number of molecules in the lower state $n_1$ is expressed in terms of the total number of molecules per $cm^{-3}$ by means of Eq. (17),

$$n_1 = n \exp(-E_1/kT_{\text{ex}})/Q_{\text{r}},$$

one finds that the absorption coefficient for dipole transitions increases approximately with $\nu^2$ at low frequencies and decreases exponentially when the energy of the lower level $E_1$ becomes comparable to $kT_{\text{ex}}$.

As an illustration of Eq. (29), the peak absorption coefficients $\alpha'_{\text{max}}$ for the rotational transitions $(J+1, K=0) \leftarrow (J, K=0)$ of methylcyanide, $CH_3CN$, have been plotted, as shown in Fig. 15. Assuming thermodynamic equilibrium with $T_{\text{ex}} = 150\,°K$, the strongest rotational lines occur for the $J = 23-22$, $K = 0$ transition at a wavelength $\lambda = 0.7$ mm. Each $J$ transition is split by centrifugal distortion effects into $(J+1)$ different $K$ components. This $K$ structure is explicitly given for the $J = 6 \leftarrow 5$ transition and it may be noticed that, because of the different nuclear statistical weights, the strongest line of this transition is the $K = 3$ line.

A comparison may be made between the calculated $J = 6-5$ transition of $CH_3CN$ on the one hand and the interstellar and laboratory spectra on the other shown in Figs. 15 and 16. The latter figure presents the emission lines observed by Solomon *et al.* in Sgr B2 and the laboratory spectrum of this transition ob-

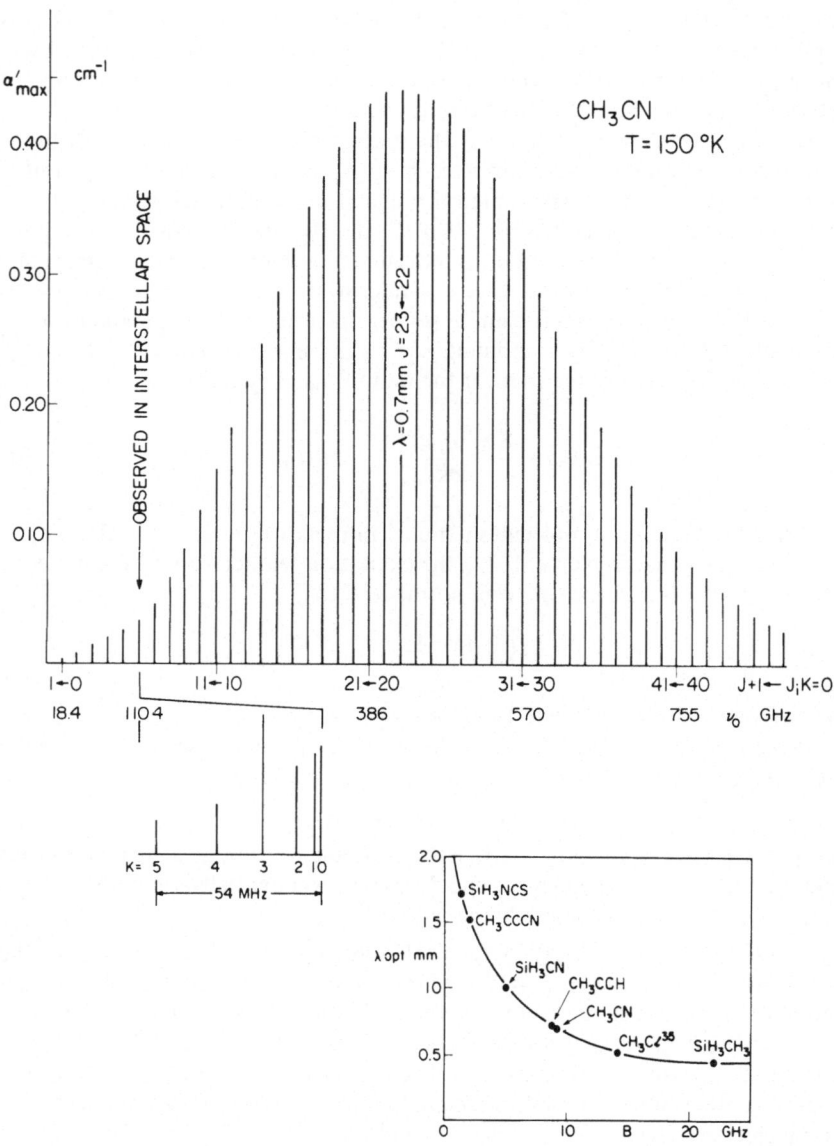

Fig. 15. Plots of the peak absorption coefficients of the $K = 0$ lines of different $J$ transitions for methylcyanide. The variation of intensity for the different $K$ components is shown for the $J = 6 - 5$ transition. The inset gives the wavelength region in which the strongest rotational lines occur for various symmetric top molecules. The excitation temperature is assumed to be 150 °K

CH₃CN

Interstellar
space

$J = 6 \longrightarrow 5$

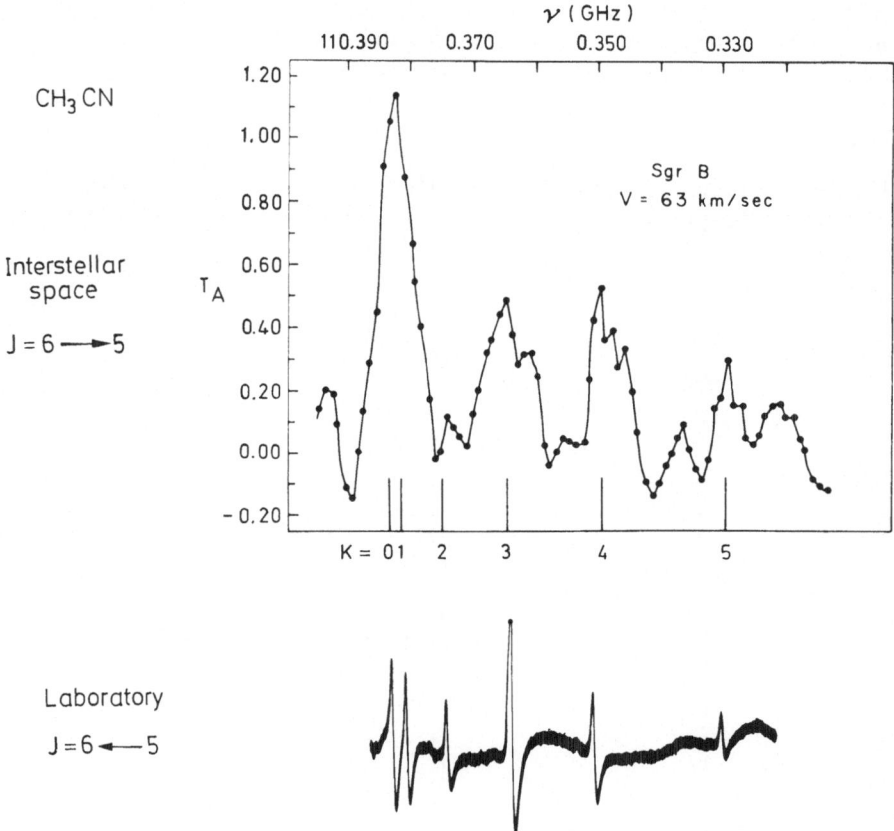

Laboratory

$J = 6 \longleftarrow 5$

Fig. 16. Comparison of the interstellar emission spectrum of CH₃CN with the laboratory absorption spectrum of the $J = 6 - 5$ transition. The interstellar emission spectrum is seen in the direction of Sagittarius B2, and shows the various values of $K$ (Solomon *et al.*, 1971). The line shapes of the laboratory spectra resemble a first derivative

served in absorption by one of us (G.W.). The observed line intensities for both cases are compared in Section III. F.

For some symmetric-top molecules (with different rotational constants B) of astrophysical interest, the calculated optimum spectral region of the strongest absorption lines ($T_{ex} = 150 °K$) lies between 0.5 mm $< \lambda_{opt} <$ 2.0 mm. These values shift towards longer wavelengths for lower excitation temperatures. Similarly, as is shown in the inset of Fig. 15, heavier molecules characterized by a smaller rotational constant B have their optimum absorption at longer wavelengths.

The energy levels of NH₃, H₂CO, and H₂O will be discussed in somewhat greater detail. In the first place they may be considered representative of the en-

43

G. Winnewisser, P. G. Mezger and H.-D. Breuer

ergy levels of symmetric and asymmetric tops and secondly, they serve to demonstrate some of the interesting features of interstellar molecular spectra related to the excitation and de-excitation of molecular transitions.

## 1. $NH_3$-molecule

The first few rotation-inversion levels of $NH_3$ are presented in Fig. 17. They are specified by the quantum numbers $J$, the total rotational angular momentum,

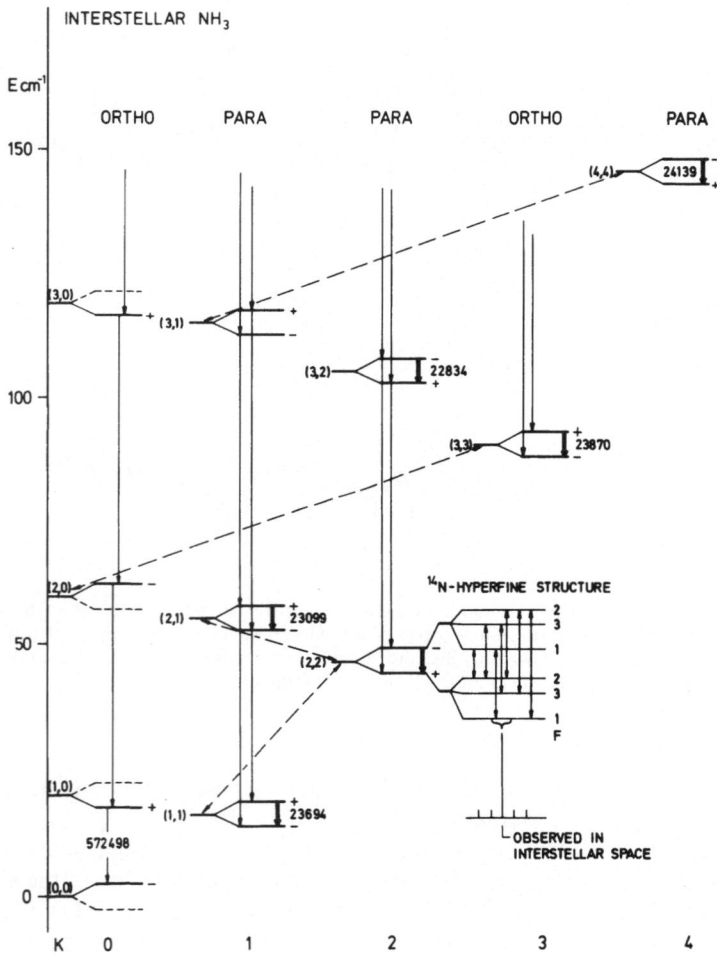

Fig. 17. Energy levels of the rotation-inversion spectrum of ammonia. The quantum numbers $(J,K)$ are given for each level. The heavy arrows indicate the inversion transitions detected in interstellar space and their frequencies in MHz. Thin arrows indicate the rotation-inversion transitions located in the submillimeter wave region. Dashed arrows indicate some collision induced transitions

44

and $K$, its projection along the symmetry axis, and the parity. Because of the inversion motion of the nitrogen atom through the plane of the hydrogen atoms, each level is split into two inversion levels. Transitions between these levels give rise to the well-known inversion spectrum in the microwave region near 23 GHz.

Fig. 17 indicates for the (2,2) transition the nuclear quadrupole hyperfine structure ($^{14}$N nucleus) which is readily observed in the laboratory not only for this but for all transitions. Under interstellar conditions, however, only the strong $\Delta F = 0$ transitions can be seen. The interstellar microwave transitions corresponding to the inversion doublets $(J, K) = (1,1), (2,2), (3,3), (4,4)$, and (6,6) were detected by Cheung et. al. (1968). Since these initial observations, emission from the energy levels $(J, K) = (2,1)$ and (3,2) has also been observed (Zuckerman et al. 1971). It may be noted that the strong rotation-inversion transitions of NH$_3$ fall in the submillimeter-wave region ($> 500$ GHz) and have, therefore, not yet been detected in interstellar space. It may be further noted that the lowest energy level in each $K$ stack is metastable since transitions to lower lying levels are forbidden by the $\Delta K = 0$ dipole selection rule.

Furthermore, the NH$_3$ molecules consist of two nuclear spin modifications of total spin 3/2 (parallel) and 1/2 (antiparallel). The Fermi statistics of the three hydrogen nuclei divide the NH$_3$ rotational energy levels into an ortho- and para-species, respectively, depending on whether $K$ is a multiple of 3 or not. Transitions between the two modifications are strongly forbidden. As a further consequence, for $K = 0$ only alternating levels occur.

## 2. H$_2$CO-molecule

The energy levels of formaldehyde, H$_2$CO, like those of any slightly asymmetric rotor, differ from the energy levels of a symmetric top in that every level with $K \neq 0$ shows a splitting caused by the asymmetry of the molecule. This is generally referred to as $K$-doubling and produces $(2J+1)$ distinct rotational sub-levels for each value of $J$. With increasing asymmetry, the $K$-doubling is increased, and finally the correspondence between the levels of a symmetric- and an asymmetric-top molecule is lost. The asymmetry is proportional to the difference in the moments of inertia around the b and c axes. Transitions between the $K$-doublets ($\Delta J = 0$, or Q – branch transitions) occur for a large number of molecules in the cm-wave region, whereas transitions between different $J$ levels (R-branch transitions $\Delta J = +1$) often lie in the millimeter-wave region. This is illustrated in Fig. 18, where the lowest energy levels of H$_2$CO are shown together with the allowed $\Delta K_a = 0$ rotational transitions. The energy levels are commonly specified by the values of $J$, $K_a$ and $K_c$ in the form $J_{K_a K_c}$, where $K_a$ is the projection of the total quantum number $J$ along the a axis and $K_c$ that along the c-axis. Just as the ammonia molecules are classified into ortho-and para-species, formaldehyde molecules exist in these two forms as well, now depending on whether the quantum number $K_a$ is even or odd. It may be noted that any asymmetric rotor molecule whose geometry allows an interchange of identical nuclei (here hydrogen) by a 180° rotation about one of the principal molecular axes exists in both ortho- and para-forms.

45

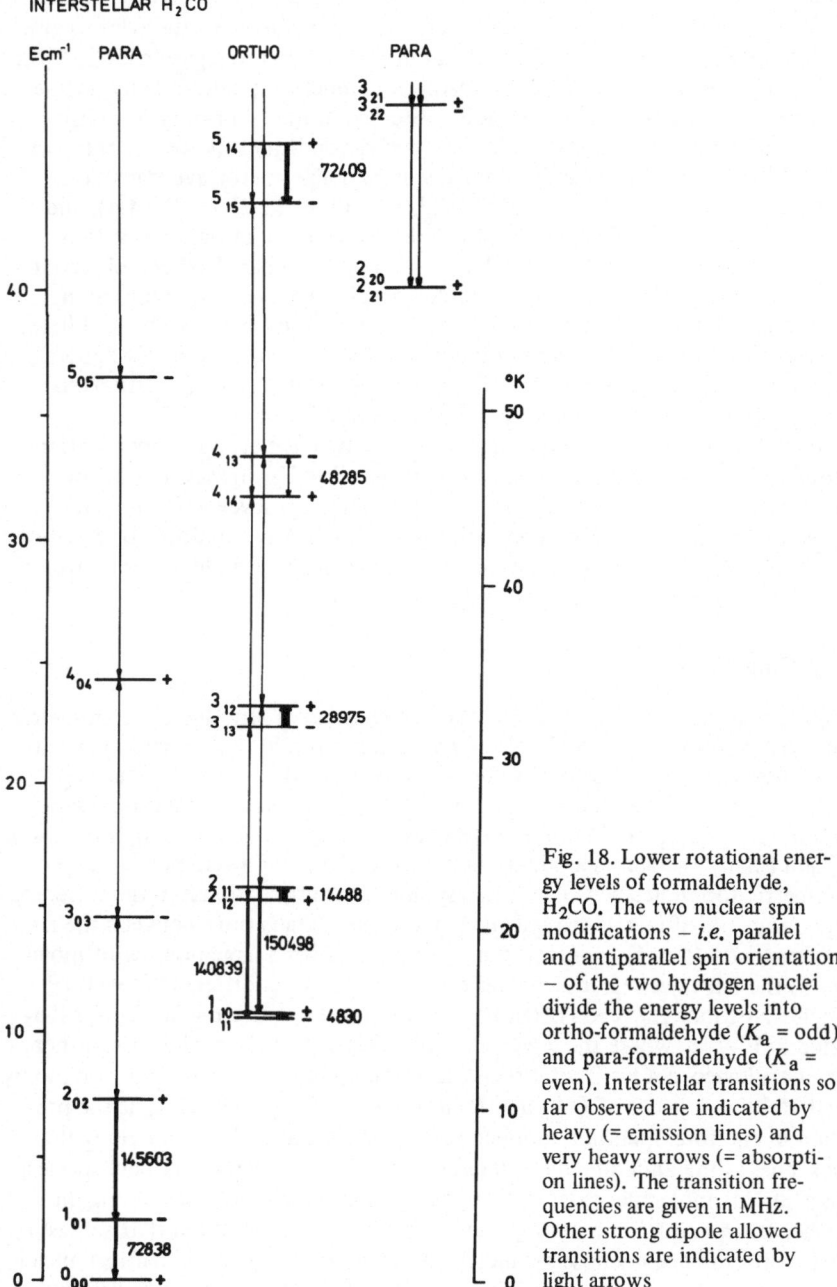

Fig. 18. Lower rotational energy levels of formaldehyde, $H_2CO$. The two nuclear spin modifications – *i.e.* parallel and antiparallel spin orientation – of the two hydrogen nuclei divide the energy levels into ortho-formaldehyde ($K_a$ = odd) and para-formaldehyde ($K_a$ = even). Interstellar transitions so far observed are indicated by heavy (= emission lines) and very heavy arrows (= absorption lines). The transition frequencies are given in MHz. Other strong dipole allowed transitions are indicated by light arrows

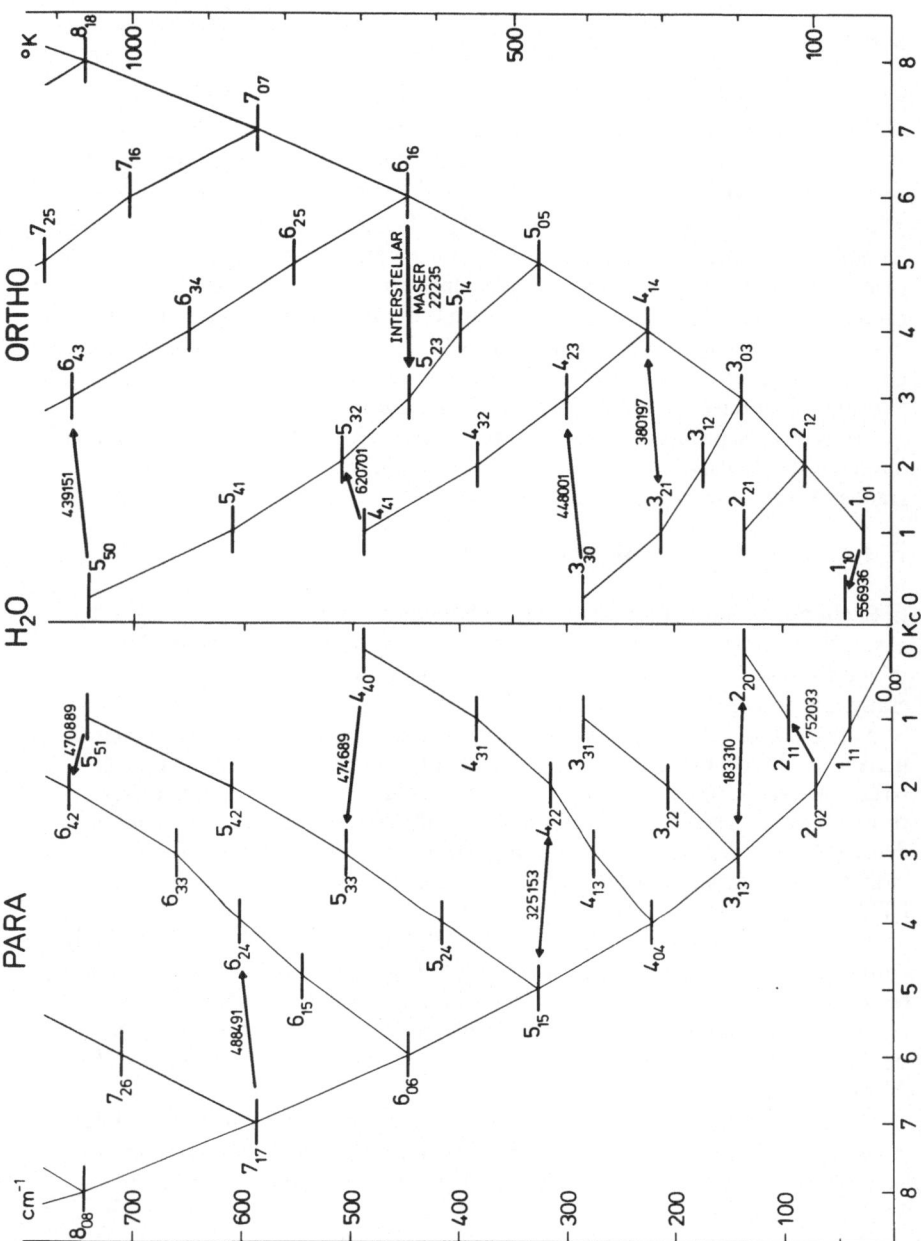

Fig. 19. Rotational energy levels of $H_2O$ divided into para- and ortho-species and sorted according to $K_c$. The strongest dipole transitions are indicated by thin lines. Heavy arrows indicate some microwave and millimeter wave transitions, together with the transition frequency in MHz. Double arrows indicate transitions of potential astrophysical interest which may appear in maser emission. It may be noted that these four transitions are the only way out of the series of lines with $K_c = J$

## 3. $H_2O$-molecule

Similarly water may be classified as either ortho-water or para-water. Its rotational energy levels are plotted in Fig. 19, sorted according to the value of $K_c$ and its nuclear spin modifications. Here the $K_a K_c = eo$ (e = even, o = odd), and the oe energy level set forms the ortho levels, and the ee and oo set the para-levels.
This results in the lowest state of ortho-$H_2O$ being the $1_{01}$ level which is about 24 cm$^{-1}$ above the ground state level $0_{00}$ of the para-form. The pure rotational spectrum of water vapor is very complex and most of its strong transitions are found to lie in the far infrared region, with only one low $J$ transition occuring in the centimeter wave region ($6_{16}-5_{23}$ = 22 235.07985 MHz) and only one in the entire millimeter-wave region ($3_{13}-2_{20}$ = 183 310.117 MHz). Throughout the submillimeter-wave region ($<$ 800 GHz) presently accessible to laboratory spectroscopy by means of microwave techniques 13 more transitions have been observed (De Lucia *et al.*, 1972), some of which have been entered on the diagram and may have astrophysical significance.

### E. Interpretation of Observed Lines

In Section B we have discussed how the basic quantities of line emission and absorption, the excitation temperature $T_{ex}$ and optical depth $\tau$ can be determined from observations. Energies required for rotational excitation are generally low enough ($<$ 200 cm$^{-1}$) so that the rotational levels are expected to be populated even at the very low kinetic temperatures of the interstellar molecular clouds. On the other hand, with a few exceptions such as $H_2O$ and $NH_3$, one may assume that only the lowest energy levels of interstellar molecules are populated. Under these conditions the observable fractional column density $N_1$ may not deviate appreciably from the total column density $N$ of a molecule, which can be computed by means of Eq. (17) on the assumption of LTE.
However, the assumption of LTE for molecules in interstellar space is highly doubtful. In fact— as will be shown in the following Section — there are a number of observational results which clearly indicate deviations from LTE emission and absorption in interstellar space and, as soon as more interstellar molecular transitions will be observed, LTE distributions may well turn out to be the exception rather than the rule.
The population of the molecular energy levels in interstellar space is determined by microwave and infrared radiation and by collisions with various collision partners (including dust grains) having kinetic temperatures between 5 °K and $<$ 100 °K.
Whether the excitation temperature $T_{ex}$ of a particular transition $u - l$ is dominated by the microwave radiation field or by the kinetic gas temperature depends on the ratio of collisionally induced transitions to spontaneous transitions, as outlined by Eq. (21).
For typical microwave transitions ($\lambda \sim$ 1 cm and dipole moment $\mu \sim$ 1 Debye = $10^{-18}$ electrostatic units), spontaneous and induced emission rates are

of the order of $10^{+3}$ to $10^{+12}$ seconds, whereas for typical interstellar conditions (density $n \sim 10^3/cm^3$, $v \sim 10^4$ cm/sec) the transition rates for collisions are of the order of about $10^{+7}$ sec or roughly one year. Thus the transition rates of the two effects are similar and since their magnitudes cannot easily be given a priori, the population distribution within a pair of energy levels cannot generally be predicted. The fact, however, that many molecular lines are seen in emission indicates that the excitation mechanism at work must be faster than the radiative transition rates. If collisions are responsible for the excitation, it becomes possible to give a lower limit to the density of the collision partner by assuming that the probability of spontaneous emission $A_{ul}$ is small compared to the probability of a collisionally induced transition $C_{ul} = n < \sigma_{ul} v >$. This yields a lower limit for the density of the emitting cloud,

$$n > \frac{64\pi^4 |\mu|^2}{3h < \sigma_{ul} v >} \cdot \frac{1}{\lambda^3} .$$  (30)

Here $\sigma_{ul}$ is the cross section for a collisionally induced transition and $v$ is the thermal velocity of the colliding particle, $< \sigma v >$ is the average value of $\sigma v$ for a Maxwellian velocity distribution. Assuming a typical dipole moment of 1 Debye, $\sigma = 10^{-15}$ cm$^{-2}$, $v = 5 \times 10^4$ cm/s, one obtains the density $n \approx 10^3/\lambda^3$. Thus for the detection of an emission line in the centimeter-wave region ($\lambda = 1$ cm) the density within the cloud is expected to be of the order of $10^3$ cm$^{-3}$. On the other hand, a detection of a millimeter-wave transition in emission at $\lambda = 1$ mm requires densities of the order of $10^5$ to $10^6$ particles/cm$^3$.

As a consequence, one finds that molecular lines located in the centimeter-wave region, particularly of molecules with medium dipole moments, are ideal thermometers for the investigation of interstellar clouds with medium gas densities $10^3 - 10^4$ cm$^{-3}$. It may be noted, however, that CO, despite its millimeter-wave transition, belongs to this class of molecules which is thermalized at relatively low densities due to its small dipole moment ($\mu = 0.11$ Debye). Furthermore, at medium cloud densities of $10^2 - 10^3$ neutral particles, transitions induced by free electrons may be possible although the electron densities are low (see Fig. 8b). Whether this is an important mechanism in exciting molecular rotational lines is not clear. Millimeter-wave transitions of molecules with large dipole moments such as $H_2CO$, CS, HCN on the other hand are ideal for probing denser clouds ($10^6$ cm$^{-3}$). In the case of CO, it seems likely that the kinetic temperature $T_k$ is close to the excitation temperature $T_{ex}$ of the molecules which, in the direction of the central part of the Orion nebula reaches a peak temperature of 80 °K (Penzias et al., 1971). Furthermore, the temperature decreases smoothly with increasing distance away from the dense central core. In "dark clouds" (see Section II. D) it is lowered to about 2 degrees above the isotropic background radiation. As the molecular energy level schemes increase in complexity (for polyatomic molecules) the interpretation of $T_{ex}$ in terms of the physical properties of the clouds becomes considerably more difficult and uncertain. However, with every new molecular line detected specific and independent information on the population distribution is obtained.

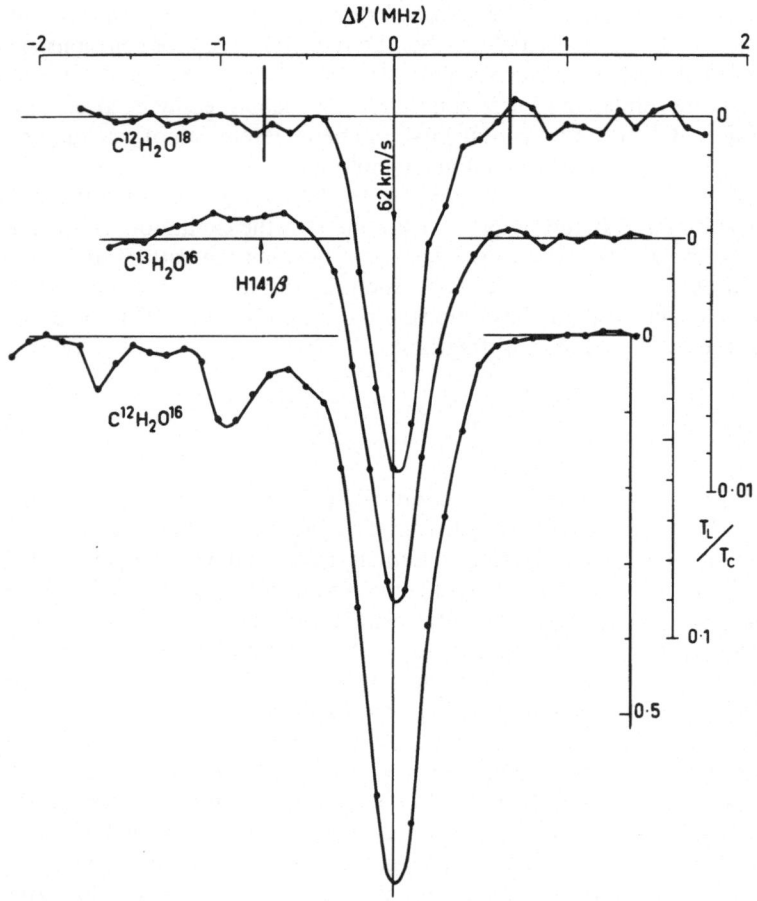

Fig. 20. Spectra of the $1_{10} - 1_{11}$ transitions for three isotopic species of formaldehyde seen in absorption against the radio source Sagittarius B2. The ordinate is the ratio of line/continuum antenna temperatures. The profiles are aligned at a velocity of 62 km/sec with respect to LSR. The center of the H141$\beta$ recombination line for a velocity of 62 km/sec is indicated on the $H_2^{13}C^{16}O$ profile (from Gardner *et al.*, 1971)

Fig. 20 shows the observed interstellar molecular lines of various isotopic species of formaldehyde, $H_2CO$, as detected by Gardner *et al.*, 1971. This particular line, the lowest asymmetry-doublet transition $1_{10} - 1_{11}$, is seen in absorption in the continuum radiation of the strong radio source Sgr B2, which is located behind the molecular gas cloud. Frequency is plotted along the abscissa and the ordinate is intensity, expressed in the ratio of line-to-continuum antenna temperatures. For all three formaldehyde isotopes the continuum temperature is $T_c \gg T_{bb} - T_{ex}$. This is the case because the formaldehyde molecules are in approximate equilibrium with the microwave background radiation and therefore $T_{bb} - T_{ex} \simeq 0$. Thus according to Eq. (25c) the plotted quantity $T_L/T_c \sim (1 - e^{-\tau})$ for $H_2{}^{12}C^{16}O$, whereas for $H_2{}^{13}CO$ and $H_2C^{18}O$ this quantity becomes proportional to the optical depth of the line (Eq. 26), from which the relative isotopic abundance is estimated. The emitting clouds have a radial velocity of $v_R$ of +62 km/sec[d]. This radial velocity introduces a line shift with respect to its rest frequency $v_L$ of $\Delta v = v - v_L$

$$v_R = - \frac{\Delta v}{v_L} c \tag{31}$$

The minus sign is due to a convention in astronomy, *i.e.* a "redshift" ($\Delta v < 0$) corresponds to a recession of the observed object. Radial velocities play a very important role in astrophysics because they allow estimates of the (kinematic) distance of the emitting object.

For optical depth $\tau \ll 1$ the observed interstellar molecular lines usually have a gaussian shape. This is to be expected, since collisional broadening, which causes the Lorentzian line profile, should be negligible and become important only at gas densities $10^{12} - 10^{14}$ cm$^{-3}$. If thermal motions of the molecules were the only source of line broadening the line half power width (*i.e.* the width between half power points) would be given by

$$\Delta v_D = \frac{2v_L}{c} \left[ ln \ \frac{2kT_k}{m} \right]^{1/2} \tag{32}$$

with $m$ the mass of the emitting molecule. However, in nearly all cases one observes linewidths which are considerably broader than one would expect on the basis of independent determinations of the kinetic temperature of the emitting or absorbing molecules. In this case one attributes the additional broadening to turbulent velocities, characterized by the rms turbulent velocity $<v_t>_{rms}$. Then the line width becomes

$$\Delta v_D = \frac{2v_L}{c} \left[ ln \ \{ \frac{2kT_k}{m} + \frac{2}{3} <v^2{}_t> \} \right]^{1/2} \tag{33}$$

However, as a result of the relatively large beamwidth of radiotelescopes the apparent turbulent velocities of molecular clouds are partly the result of large-scale systematic motions, such as rotation.

---

[d] Radial velocities are given with respect to the Local Standard of Rest (LSR) which accounts for both the motion of the earth and the local group of stars.

## F. Observed Deviations from Local Thermodynamical Equilibrium

There are some particularly interesting phenomena encountered in the observation of interstellar molecular spectra: the experimental data show that the rotational distribution is, in most cases, anomalous. There is no common single excitation temperature which can describe the interstellar population distribution over the energy level manifold in terms of a Boltzmann distribution.
From the preceding discussions it is evident that at least four different temperatures have to be considered which under laboratory conditions are all equal: the excitation temperature $T_{ex}$ of the molecule, defined by the relative populations of the levels, the kinetic temperature $T_k$, corresponding to the Maxwellian velocity distribution of the gas particles, the radiation temperature $T_{rad}$, assuming a (in some cases diluted) black body radiation distribution, and the grain temperature $T_{dust}$. With no thermodynamic equilibrium established, as is common in interstellar space, none of these temperatures are equal. These non-equilibium conditions are likely to be caused in part by the delicate balance between the various mechanisms of excitation and de-excitation of molecular energy levels by particle collisions and radiative transitions, and in part by the molecule formation process itself. Table 7 summarizes some of the known distribution anomalies. The non-equilibrium between para- and ortho-ammonia, the very low temperature of formaldehyde, and the interstellar OH and $H_2O$ masers are some of the more spectacular examples.

An example of non-equilibrium has been observed in the interstellar spectrum of methyl cyanide. In comparing the $K$-structure of the $J = 6 - 5$ transition of interstallar $CH_3CN$ detected in the direction of Sgr B2 (Solomon et al., 1971) with the laboratory spectrum (see Fig. 16) it becomes evident that the $K = 2$ line seems to be missing in the interstellar spectrum. This can only be interpreted by concluding that for some unknown reason the $K = 2$ energy levels are considerably depopulated. No explanation has been given so far; however, this effect should be substantiated by the detection of the connecting higher and lower $J$ rotational transitions.

*Note Added in Proof.* Recent observations by Solomon et al. 1973 show that the $K = 2$ line is present at or near its expected intensity. Thus this puzzling result of the first detection has been in error.

In contrast to the $CH_3CN$ situation, the spectra of interstellar ammonia give considerable insight into excitation and de-excitation mechanisms. From the observed intensities of the interstellar ammonia lines it has been derived that the excitation temperature $T_{12}$, determined from the relative intensities of the (1,1) and the (2,2) lines, is notably lower than the excitation temperature $T_{13}$ determined from the intensities of the (1,1) and (3,3) lines. Thus the (3,3) level shows an excess population over the (1,1), (2,2) levels. In other words, ortho-ammonia is not in equilibrium with para-ammonia. However, a more detailed study of the two para-ammonia levels (1,1) and (2,2) also reveals that their relative populations are not given by a simple Boltzmann factor for each of them. The (1,1) level has population in excess over the Boltzmann distribu-

Table 7. Observed non-Boltzmann distributions within the rotational levels of several interstellar molecules

| Molecule | Anomalous spectral features | Excitation temperature |
|---|---|---|
| OH | Maser emission | $T_{ex} < 0$ |
| $H_2O$ | Maser emission | $T_{ex} < 0;$ |
| $NH_3$ | Anomalous distribution for para-$(K = 1, 2, \ldots)$ vs ortho $(K = 3, \ldots)$ levels | $T_{ex}(\text{para}) \approx 35\,°K$ $T_{ex}(\text{ortho}) \approx 50\,°K$ |
| $H_2CO$ | Anomalous absorption of $1_{10} - 1_{11}$ level | $T_{ex} \approx 0.8 - 1.8\,°K$ |
| $CH_3OH$ | Anomalous distribution for $A, E_1, E_2$ levels | $(T_{ex} < 0\,?)$ |

tion. This type of inequilibrium could be explained by collisions of $NH_3$ molecules with $H_2$ molecules. Extensive laboratory experiments by Oka and collaborators (1970) support this explanation by demonstrating clearly that the $K$ quantum number is changed in units of $3\hbar$ in $NH_3$-rare gas or $NH_3$–$H_2$ collisions. In the energy diagram of ammonia, some of the collisionally allowed transitions are indicated by dashed arrows. Thus, for example, collisional transitions can transfer molecules from the (2,2) level to the (1,1) — as well as to the (2,1) level. Molecules in the (2,1) level decay by radiation very quickly into the (1,1) level, giving it the observed excess population. Oka's experiments show further that transitions between ortho- and para-ammonia are "forbidden" even in collisions. Thus the observed excess population of the interstellar ortho-level cannot be explained by the collisional mechanism. Furthermore, it can be shown (Oka *et al.*, 1971) that $\Delta K = 3$ radiative transitions are weakly allowed, although only between levels of the same species. By this mechanism the (3,3) level can decay to the (2,0) level in a time span of about 40 years. At low temperatures interstellar ortho-ammonia may be produced more efficiently than para-ammonia.

Recently, however, emission from the (2,1) and (3,2) levels has been observed (Zuckerman *et al.*, 1971). These levels radiatively decay in a very short time to the lower levels (see energy diagram). Thus collisional excitation of these levels would require molecular densities of probably $10^7$ of $10^8$ cm$^{-3}$, so that these transitions should be ideal for investigating the dense cores of black clouds.

Another striking example of non-equilibrium conditions in interstellar space is the observation of the anomalous absorption of $H_2CO$ in various $K$-type levels. The $\lambda 6$ cm rotational transition $1_{10} - 1_{11}$ has been found in absorption in many galactic continuum sources (*i.e.* HII regions and supernovae remnants), even when their continuum temperatures are only a few degrees, indicating that the excitation temperature of the $\lambda 6$ cm line must be very low, close to the 2.7 °K of the microwave background. It was, however, completely unexpected that the line could still be seen in absorption in many dark nebulae where no continuum background source is located beyond the dark cloud. According to Eq. 26 this result means that the excitation temperature of this line must be lower than 2.7 °K (Palmer *et al.*, 1969). This phenomenon is referred to as anomalous absorption of the microwave background radiation by the $H_2CO$ $\lambda6$ cm line. Formaldehyde has thus become the second molecule next to CN to produce clear evidence of the existence of a microwave background radiation.

In the absence of any interactions, the population of the formaldehyde levels would rather quickly ($\sim 10^6$ sec) reach equilibrium with the 2.7 °K radiation. Anomalous absorption, however, implies that a widespread and fast-pumping mechanism must exist which cools the levels below 2.7 °K. Several mechanisms have been proposed to explain this low excitation temperature: i) in terms of collisions with atomic or molecular hydrogen followed by radiative de-excitation (Townes and Cheung, 1969); ii) small deviations in the microwave background radiation near 2 mm (Solomon and Thaddeus, 1971); iii) molecules formed in higher rotational levels decay down to lower rotational levels undergoing "adiabatic cooling" as they follow the radiative dipole selection rules (Oka, 1970); iv) by infrared pumping caused by shock-heated layers in gravitationally unstable dust clouds (Litvak, 1970).

Although observational and theoretical evidence seems to favour some of the proposed mechanisms more than others it still remains open as to which one produces the refrigeration effect. A more direct selection between the proposed models may become possible from the detection of other $K$-type doublets of formaldehyde in anomalous absorption and also of similar transitions of structurally related molecules like $H_2CS$, $H_2C_2O$.

In contrast to this "anti-maser" action in formaldehyde, $H_2O$ and OH are observed in maser emission. Collisional or radiative pumping is thought to maintain the population inversion between the two levels. As photons pass through the cloud, they are amplified by stimulated emission of radiation. The maser emission of $H_2O$ is possibly the most unusual of the observed anomalies, both from an astrophysical and a spectroscopic point of view. This is not the place to discuss details of the various models suggested, we would refer to concentrate on some of the general features observed in the maser emission spectra.

One of the striking features of interstellar maser emission is the enormous intensity the maser lines have. In the case of water, the brightness temperature for the source W49 reaches about $10^{15}$ °K. Furthermore, the line widths of the observed lines are extremely narrow, typically only a few ten of kHz. Both properties, intense and narrow emission lines, are intrinsic indications of maser emission. It has been found that the angular size of all interstellar maser sources is very small, *i.e.* much smaller than the spatial resolution obtained with large single dish radiotelescopes. From long baseline interferometry, however, an upper limit has been placed on the apparent source size of about 0.002 seconds of arc (for W49 = 0.0003", Orion = 0.001") (Hills *et al.*, 1972), which, for example, at the distance of Orion, 450 pc, makes this particular water vapor source about 1/2 AU in size. This is comparable with the diameter of a red giant star. Furthermore, the maser emission of OH and $H_2O$ does not occur at a single frequency but consists generally of many peaks corresponding to different Doppler shifts. For $H_2O$, the Doppler shifts in W49 range from −180 km/sec to +210 km/sec. Other pronounced spectral features include the observed change in line intensities in a time scale as short as a few days and the occurrence of polarization which may reach about 20% linear polarization for $H_2O$ but is often much higher for OH sources, where in some cases 100% circular polarization has been observed, presumably caused by the difference in the Zeeman effect of the two molecules, *i.e.* the greater sensitivity (of the order of $\sim 10^4$) of the OH molecule to a magnetic field.

From the spectroscopic point of view the negative temperature in $H_2O$, and thus the occurrence of maser emission, is surprising, since the $6_{16} - 5_{23}$ transition of ortho-water arises from rotational energy levels which are about 450 cm$^{-1}$ above the ground state level, usually too high to show any appreciable populations. Furthermore, transitions from both levels down to lower levels are strongly allowed, as indicated in Fig. 19 along with other strong transitions due to spontaneous emission. If one follows these paths, one finds that the water molecules are funneled by radiative transitions preferentially into the $J = K_c$ energy levels (Winnewisser, 1972a) to which the $6_{16}$ level belongs. Thus, once an excitation mechanism efficiently pumps the molecules into higher energy levels ($J \geqslant 10$), a population inversion between the two levels may occur. It is interesting to note, as Oka (1971) has pointed out, that the microwave transition which is observed in interstellar space, is the first way out of the strongly populated $J_{0,J}$ and $J_{1,J}$ levels. Similar considerations hold for para-water whose $3_{13} - 2_{20}$ transition at 183.3 GHz is, therefore, of particular interest. On the basis of this model we might mention that $H_2S$ has a similar energy level diagram. Two transitions could be expected to show maser emission: $4_{14} - 3_{21} = 204.140$ GHz of ortho-$H_2S$ and $3_{13} - 2_{20} = 392.618$ GHz of para-$H_2S$.

The fact that the emission from $H_2O$ and OH comes from many different but very intense spots $(1 - 100$ AU in linear dimensions) separated by distances of about 10,000 AU lends support to the suggestion that this radiation is emitted from massive protostars in their pre-Main Sequence adiabatic contraction state (Mezger and Robinson, 1968; Mezger, 1971).

## G. Conclusion and Summary

The very existence of molecules in interstellar space has provided considerable new insight into the interstellar chemistry of our Galaxy. By providing birth and certainly shelter to molecules like HCN, $H_2O$, $NH_3$ and $H_2CO$, known to be important in reactions which synthesize amino acids, the interstellar environment is not nearly as hostile as had originally been assumed.

Since the detection of interstellar $NH_3$, some 28 different molecules have been identified by means of their radio lines. A steadily growing number of

55

transitions in the radio region remains unidentified, since they cannot be assigned to spectra of presently known molecules. Most of the molecules especially the complex organic ones, are found predominantly in dark and black clouds which are dense and cool condensations of the interstellar gas out of which stars form. These regions cannot be studied by optical means nor by the $\lambda21$ cm atomic hydrogen line, but only by means of molecular lines. The interpretation of observed molecular spectra allows us to learn more about the physical conditions, the chemical pathways of molecule formation (see Section IV), and the relative isotopic element abundances. So far the isotopes D, $^{13}$C, $^{15}$N, $^{18}$O and possibly $^{34}$S (Zuckerman, 1972) have been found in various molecular species.

In contrast to radio observations by which dense clouds are investigated, optical observations yield information on dilute clouds only. Both methods yield column densities and from these values space densities can be estimated. Such results show that CO is probably next to $H_2$ the most abundant molecule, whereas OH and $H_2CO$ though not the most abundant molecules seem to be most widely distributed. Some of the more complex molecules are just barely detectable by present radio techniques. Improved experimental techniques will be paralleled by the detection of other molecules.

The location of an observed molecular radio transition in its energy level scheme and its measured interstellar intensity contain important information concerning the physical state of the molecular cloud in which the transition is observed. It will therefore be an important task for future interstellar molecular research to observe and measure as many transitions of any one molecule in any one particular cloud. Doppler shifts, *i.e.* the difference in frequency between the rest frequency (known from laboratory measurements) and the observed interstellar line frequencies, provide information on the large scale motion of the molecular clouds while the linewidths indicate the turbulence within the clouds.

Interstellar molecular lines are seen in absorption, emission and maser emission. The fact that all millimeter-wave transitions are seen in emission puts a lower limit on the density of the gas ($n \gtrsim 10^3$ particles cm$^{-3}$). If different molecular lines can be observed for one given species then the intensities of the transitions yield information as to whether the molecule is in LTE or not. In most cases where more than one rotational transition for one molecule could be observed these show more or less strong deviations from LTE. Striking deviations from LTE are the maser emission of interstellar OH and $H_2O$ originating from very localized sources ($\sim 1/2$ AU) suggesting very high densities ($\sim 10^9$ particles cm$^{-3}$), and the anomalous absorption of the $H_2CO$ $\lambda6$ cm line which furnishes evidence of the microwave background radiation at that wavelength.

## IV. Formation and Destruction of Molecules in Interstellar Space

### A. Introduction

The existence of a large number of molecules and their concentrations in interstellar space suggest that the interstellar medium is quite heterogeneous. Although the chemical processes are likely to be different in the various regions of space, it is possible to consider some common features of the life cyle of an interstellar molecule. The important processes are the formation, the relaxation, and finally the destruction of a molecule. Formation and destruction are not well understood processes.

Possibly best understood is the dissociation of a molecule as part of the destruction process. All molecules observed to date in interstellar space can be dissociated by absorption of UV photons of wavelengths $\lambda > 912\,\text{Å}$ (see Fig. 7). Exchange reactions and adsorption on dust grains also contribute to the loss of a particular molecule. In fact the lifetime of these molecules is less than 100 years if they are not protected against the UV radiation, but does increase by several orders of magnitude in areas of strong shielding. The relaxation processes which take place during the lifetime of a molecule are important for the understanding of the observational data (see Section III) and in particular of the anomalous population distributions which most likely are influenced by the formation process as well. The formation of molecules is the least understood of the three processes, and it is in this area where new ideas are needed. Although various formation mechanisms have been discussed two main processes seem to contribute to molecule formation. i) Gas phase reactions caused by binary collisions in low and medium density clouds ($n_{\text{H}_2} < 10^3$ particles cm$^{-3}$) produce diatomic and possibly simple polyatomic molecules. Ion-neutral gas phase reactions which are known to be very fast may be important for clouds with densities $10^4 \leqslant n_{\text{H}_2} \leqslant 10^6$ cm$^{-3}$, whereas ii) surface reactions on dust grains are assumed to be operative over a very wide range of cloud densities forming diatomic as well as polyatomic molecules.

Interstellar molecules are observed in the cold component of the interstellar gas which, together with the short lifetime of the molecules in unshielded regions of space, shows that formation in stellar atmospheres or protostellar nebulae can not be their principal formation mechanisms provided the molecules are not formed simultaneously with the dust and then blown out into the interstellar medium.

The wide variety of interstellar molecules detected so far in our Galaxy (see Table 6) are composed of the most abundant chemically reactive elements, *i.e.* H, C, N, O, Si and S. The selection of detected molecules is influenced by molecular and observational considerations: i) the molecules must be polar ii) they must have sufficient vapor pressure for their laboratory spectra to be known, iii) of the known spectra, only the most intense transitions can be expected to be observable in interstellar space, and iv) the frequencies of these transitions have to be located within the Earth's atmospheric windows Only molecules which satisfy these conditions are amenable to radio techniques.

57

This may have excluded from detection so far molecules containing the elements Fe and Mg which are of comparable abundance to Si and S. Furthermore, molecules which are expected to be abundant in interstellar space, such as HCCH, $CH_4$, $CO_2$ cannot at all or only with great difficulty be detected in the microwave region.

Nevertheless the present observational evidence within its many limitations indicates that closed-shell molecules are favored over open-shell molecules (*i.e.* free radicals) at least in the molecule rich dark clouds seen in the direction of Orion, W51, Sgr A and Sgr B2. In the latter two sources many stable and rather complex organic molecules and molecular transients, which under laboratory conditions are short-lived molecules, have been found in comparatively high abundances, whereas few free radicals occur in these regions of space. It remains, however, to be stressed that important clues on possible formation mechanisms (and the possibility to differentiate between them) may be obtained in case it can be proved that molecular ions such as $HCO^+$, $CCH^-$ and others may be the carriers of these unidentified lines. Although the present list of interstellar molecules gives an incomplete picture of the chemical conditions in space, it seems that many of the observed interstellar molecules are related in that they can be formed from molecular fragments to which successive hydrogen atoms and free radicals have been attached. Following this scheme (Winnewisser, 1972) which is shown in Table 8, one starts off with the elements of the second row of the periodic table C, N, O, F or the 3rd row elements Si, P, S, Cl, and produces by hydrogen addition successive hydride radicals, ready to react with other hydrogen atoms, or with each other, or with interstellar molecules like CO, CN, CS etc. to form the observed molecules. These reactions presumably take place on grain surfaces. Hydrogen addition to the hydride radicals leads to the closed shell hydride molecules like HF, $H_2O$, $NH_3$, $CH_4$ (and HCl, $H_2S$, $PH_3$, $SiH_4$ for the third-row elements). From these the more readily detectable molecules $H_2O$, $NH_3$ and $H_2S$ have been identified, suggesting that the other saturated hydrides may be present as well in observable quantities in interstellar space. On the other hand reactions of hydride radicals with each other lead to the formation of astrophysically important, but often non-polar molecules such as HCCH, $H_2CCH_2$, $H_3CCH_3$. Correspondingly HNNH, $H_2NNH_2$ and in particular HOOH and HSSH, which are all polar, may very well be detectable in interstellar space. Surface reactions of the hydride radicals with other radicals and molecular fragments produce more complex molecules. For example cyanoacetylene, HCC–CN may well have been formed this way. All detected interstellar polyatomic molecules can be explained this way (see Table 8) and some hitherto undetected, but important ones, can be predicted to exist in interstellar space. Their observation or absence in interstellar space may then in conjunction with laboratory results shed more light on possible chemical pathways.

The scope of this section is limited to a discussion of the formation and dissociation processes of molecules in cool clouds of interstellar gas with densities $n > 10 \text{ cm}^{-3}$ and with kinetic gas temperatures $T_k > 5 \text{ °K}$ together with some laboratory work related to the formation of interstellar molecules. Chemical processes suggested to be operative in solar nebulae are briefly mentioned.

Table 8. Schematic representation of the formation of interstellar molecules (after Winnewisser, 1972)

| Detected interstellar molecules | Hydride formation | Not yet detected molecules |
|---|---|---|

Hydride formation diagram:

≡C   ≡N   =O   –F   Ne
     ≡CH  =NH  –OH  FH
          =CH₂ –NH₂ H₂O
               –CH₃ NH₃
                    CH₄

$HC{\equiv}X;$  X=N

$HCC{-}X;$  X=CN, $CH_3$

$H_2C{=}X;$  X=O, S, NH

$H_3C{-}X;$  X=CN, CCH, OH, HCO

$\underset{H}{\overset{O}{\diagdown}}C{-}X;$  X=$NH_2$, $CH_3$, OH

CO, CN, CS, SiO, OCS, HNCO

Not yet detected molecules:

} HCCH, HNNH, HOOH

} $H_2CCH_2$, $H_2NNH_2$

} $H_3CCH_3$ ($H_3SiCH_3$)

The open shell hydrides are indicated by the dashed line forming a triangle. With the exception of OH these basic free radicals have either no or only poorly known microwave spectra. Each hydrogen addition consumes one free valence whereby the radical assumes the chemical properties of the element with the next higher atomic mass number. The closed shell hydrides form the well known stable molecules in the column beginning with Ne.

G. Winnewisser, P. G. Mezger and H.-D. Breuer

## B. Formation of Diatomic Molecules and Radicals in the Gas Phase

In low to medium density clouds ($10 \leqslant n \leqslant 10^3$ particles cm$^{-3}$) radiative
association is one important mechanism for producing the diatomic radicals
CH and CH$^+$ as has been shown by Solomon and Klemperer (1972). These
primary reactions are followed by exchange reactions which produce CN, CO,
$C_2$ but not $H_2$, NO, $N_2$ and $O_2$. Thus Solomon and Klemperer were able to
quantitatively account for Herbig's (1968) optical observations made in the
direction of ξ-Ophiuchi. They made the assumptions that i) the UV radiation
field is intense enough that the major constituents of the gas are atoms or ions;
ii) The atomic species are taken to be known fractions of the hydrogen density;
iii) the hydrogen density is of the order of 50 cm$^{-3}$ and the temperature is
about 20 °K. The low temperature of interstellar space allows the occurrence
of exothermic gas phase reactions only and for the reactions to occur at an
appreciable rate they must proceed without activation energy. In addition the
low density restricts the reaction processes to binary collisions between ground-
state atomic systems, because the time between collisions is longer than the
radiative lifetime of all optical processes with the exception of hyperfine transitions.

Two types of gas phase reactions are considered for the formation of diatomic molecules: i) radiative association, ii) chemical exchange reactions. The
rate constants[e] are either taken from laboratory experiments or estimated by
comparison with similar reactions and if possible they are corrected for low
temperature conditions. Since rate constants are often not well known and
subject to major revision as new experimental data become available they are
a source of considerable uncertainty. For radiative association the reaction rate
is equal to the product of the rate of collisions and the probability that a photon
is emitted during the collision process (Bates, 1951). The general reaction scheme
is of the form

---

[e] For a reaction A + B → C + D, the rate constant $k$ is defined by the differential equation

$$-\frac{dn_A}{dt} = -\frac{dn_B}{dt} = \frac{dn_C}{dt} = \frac{dn_D}{dt} = k\, n_A\, n_B,$$

where $n_A$ and $n_B$ are the number densities of the reacting species A, B. If one of the
reaction partners (say B) is a photon, then the rate constant is defined

$$-\frac{dn_A}{dt} = \frac{dn_C}{dt} = \frac{dn_D}{dt} = k\, n_A.$$

If, for example, the formation of $H_2$ molecules is computed, all possible binary reactions between different molecules, atoms, ions, photons, etc. have to be considered.
This yields according to the above-mentioned definitions, a set of steady state equations which involve the rate constants and the densities of the various reactants. The
latter quantities are then computed under the assumption of known rate constants.

60

$$A + B \rightarrow AB + h\nu \tag{34}$$

In particular it turns out that the reaction

$$C^+ + H \rightarrow CH^+ + h\nu \tag{35}$$

is the initial step in the formation of the species $CH^+$, CH, CN, CO and $C_2$, since the radiative recombination reaction

$$C + H \rightarrow CH + h\nu \tag{36}$$

encounters several difficulties. The low abundance of neutral carbon[f] restricts this reaction to the central core of dense clouds, but more fundamentally the experimentally observed maximum of the potential curve of the relevant $B^2\Sigma^-$ state (Herzberg and Johns, 1969), which has a height of about 500 cm$^{-1}$ relative to the separated atoms, considerably reduces the molecular production rate under interstellar conditions. It is interesting to note that the formation of CH and $CH^+$ by gas phase reactions of the form

$$
\begin{aligned}
&\text{(a)} \ C + H_2 \rightarrow CH + H \\
&\text{(b)} \ C_2 + H_2 \rightarrow 2CH \\
&\text{(c)} \ C^+ + H_2 \rightarrow CH + H^+ - 3.35 \ eV \\
&\text{(d)} \ C^+ + H_2 \rightarrow CH^+ + H - 0.40 \ eV \\
&\text{(e)} \ C_2^+ + H_2 \rightarrow 2CH^+
\end{aligned}
\tag{37}
$$

which employ $H_2$ are unlikely, since they are endothermic. If, however, $H_2$ should be vibrationally excited, with the vibrational energy exceeding the endothermicity and noting that the vibrational relaxation is very slow, then reactions 37c and 37d proceed with rate coefficients of $10^{-11}$ cm$^3$ sec$^{-1}$ (Stecher and Williams, 1972), thus dominating the production by grain-surface reactions by several orders of magnitude. Similarly $H_2$ is not formed by radiative association nor by the reactions $CH + H \rightarrow C + H_2$ and $CH^+ + H \rightarrow C^+ + H_2$ which are very slow. Thus $H_2$ does not enter in the model developed by Solomon and Klemperer. Interstellar dust is believed to provide the principal mechanism for $H_2$ formation.

---

[f] The ionization limit of C is considerably lower than that of hydrogen (see Table 1); on the other hand in interstellar space the radiation density beyond of the Lyman continuum is high enough to ionize most carbon. The transition region between the CII region and the CI region is relatively sharp. CI regions occur only in the inner part of dense clouds (see Section II. D).

Table 9. Rate constants in $cm^3/$(part. sec) for radiative association (Solomon and Klemperer, 1972)

| | | |
|---|---|---|
| $CH^+$ : | $(T < 50\ ^\circ K)$ | $7 \times 10^{-17}$ |
| | $(T = 100\ ^\circ K)$ | $3,5 \times 10^{-17}$ |
| | $(T \geqslant 200\ ^\circ K)$ | $2 \times 10^{-17}$ |
| $CH$ : | $(10\ ^\circ K \leqslant T \leqslant 50\ ^\circ K)$ | $(3,0 - (T-20) \cdot 0,06) \cdot 10^{-17}$ |
| | $(T \sim 100\ ^\circ K)$ | $3 \times 10^{-18}$ |

Once CH and $CH^+$ are formed by radiative association there exist several channels by which they can be destroyed. Chemical exchange reactions which are assumed to proceed at the classical collision frequency, unless there is evidence to the contrary, are for example the primary source for destroying CH by converting it according to the reactions: $CH^+ + O \rightarrow CO + H^+$, $CH^+ + N \rightarrow CN + H^+$ and $CH^+ + e \rightarrow CH$ into the species CO, CN and CH. According to the nature of the reacting species one differentiates between ion-molecule, charge-exchange and neutral-neutral reactions. The calculated rate constants for some of the reactions are listed in Table 10.

Table 10. Rate constants in $cm^3/$(part. sec) for chemical reactions (Solomon and Klemperer, 1972)

| Reaction | | Rate constant |
|---|---|---|
| **Ion-Molecule** | | |
| $CH^+ + O$ | $\rightarrow CO^+ + H$ | $1 \times 10^{-9}$ |
| $CH^+ + O$ | $\rightarrow CO\ + H^+$ | $1 \times 10^{-9}$ |
| $CH^+ + N$ | $\rightarrow CN\ + H^+$ | $1 \times 10^{-9}$ |
| $CH^+ + C$ | $\rightarrow C_2^+ + H$ | $1 \times 10^{-9}$ |
| $CH\ + C^+$ | $\rightarrow C_2^+ + H$ | $1 \times 10^{-9}$ |
| $CH^+ + H$ | $\rightarrow C^+ + H_2$ | $7.5 \times 10^{-15} \cdot T^{5/4}$ |
| $C_2^+ + O$ | $\rightarrow CO\ + C^+$ | $1 \times 10^{-9}$ |
| $C_2^+ + N$ | $\rightarrow CN\ + C^+$ | $1 \times 10^{-9}$ |
| **Charge-Exchange** | | |
| $CO^+ + H$ | $\rightarrow CO\ + H^+$ | $1 \times 10^{-9}$ |
| $CN^+ + H$ | $\rightarrow CN\ + H^+$ | $1 \times 10^{-9}$ |
| **Neutral-Neutral** | | |
| $CH\ + O$ | $\rightarrow CO\ + H$ | $4 \times 10^{-11}$ |
| $CH\ + C$ | $\rightarrow C_2\ + H$ | $4 \times 10^{-11}$ |
| $CH\ + N$ | $\rightarrow CN\ + H$ | $4 \times 10^{-11}$ |
| $C_2\ + O$ | $\rightarrow CO\ + C$ | $3 \times 10^{-11}$ |
| $C_2\ + N$ | $\rightarrow CN\ + C$ | $3 \times 10^{-11}$ |
| $CN\ + O$ | $\rightarrow CO\ + N$ | $10^{-11}\exp(-1200/T)$ |
| $CN\ + N$ | $\rightarrow N_2\ + C$ | $1 \times 10^{-13}$ |
| $CH\ + H$ | $\rightarrow C\ + H_2$ | $1 \times 10^{-14}$ |

There exists considerable experimental evidence that the reaction $CN + N \rightarrow N_2 + C$ is several orders of magnitude faster (Kley, 1973) than the value assumed by Solomon and Klemperer and in fact it seems to be the fastest destruction mechanism for CN. In general, destruction of molecules or radicals occurs by the interaction with UV photons and by exchange reactions. The abundance of each molecular constituent is determined by equating the total formation rate to the total destruction rate. Solomon and Klemperer solved the rate equations for about 30 reactions. Table 11 compares the predicted abundances for $\zeta$ Oph. assuming $n_H = 100$ cm$^{-3}$ and $T = 20\,°$K with the observed column densities.

Table 11. Observed and predicted column densities in cm$^{-2}$ for $\xi$ Ophiuchius (Solomon and Klemperer, 1972)

|  | Observed | Predicted |
|---|---|---|
| CH | : $4.3 \cdot 10^{13}$ | $1.7 \cdot 10^{13}$ |
| CH$^+$ | : $2.6 \cdot 10^{13}$ | $1.3 \cdot 10^{13}$ |
| CN | : $8.3 \cdot 10^{12}$ | $6 \cdot 10^{12}$ |
| NH | : $8 \cdot 10^{12}$ | 0 |
| OH | : $7 \cdot 10^{13}$ | 0 |
| CO | : ? | $2.4 \cdot 10^{14}$ |
| C$_2$+C$_2$$^+$ | : ? | $7 \cdot 10^{12}$ |

For various cloud models which differ in hydrogen density, dimensions and temperature, Solomon and Klemperer derive theoretical column-densities. For the molecules considered these are in the range $10^{10}$ to $10^{15}$ cm$^{-2}$. A comparison with observations shows a general agreement considering all the limitations imposed on the observational and theoretical determinations.

## C. Molecule Formation on Grains

The formation of $H_2$ on grains is very sensitive to the residence time of the physically adsorbed H atoms on the grain surface. The residence time is the time elapsed between adsorption and desorption of atoms on the grain surface. The molecule formation depends therefore on the grain temperature. Assuming a perfectly regular surface, the H atoms would "evaporate" from the grain surface before they could form $H_2$ molecules unless the grain temperature is below a certain critical value $T_{gr}$ which in the case of a perfect surface is $T_{gr} \leqslant 13\,°$K. This critical grain temperature may be increased to $T_{gr} = 25-50\,°$K if one assumes surface irregularities which provide localized adsorption sites to which the atoms are attached more strongly by the somewhat increased adsorption energies. Since there is strong evidence that the grain temperatures in HI regions are lower than $50\,°$K Hollenbach and Salpeter conclude that every H atom which hits a grain and sticks to it leads to $H_2$ formation before desorption. For very small dust clouds with $n_H = 10$ cm$^{-3}$ and $\tau_v = 0.1$ the fraction of hydrogen in molecular form, $f$, is calculated to be

$$f = \frac{2\, n_{H_2}}{(n_H + 2\, n_{H_2})} \approx 10^{-3} \tag{38}$$

In denser clouds, however, the abundance of $H_2$ increases rapidly. With gas densities $\geqslant 10^2$ cm$^{-3}$ and column densities corresponding to an extinction in the visible of $A_v \geqslant 0.5^m$ hydrogen is expected to be mainly in the molecular form (Werner *et al.*, 1970). Similar results are obtained by de Jong (1972) who also includes low-energy cosmic radiation in his calculations which produces electron densities in dark clouds sufficient to yield $H_2$ formation via associative detachment $H^- + H \rightarrow H_2 + e$ proceeding with a rate constant $k \sim 1.3 \times 10^{-9}$ cm$^3$ sec$^{-1}$ at 300 °K. This reaction, however, is only of importance if formation of $H_2$ molecules on grains is for one reason or another inefficient.

The difficulty encountered with physically adsorbed atoms on perfect surfaces, namely the much too short residence time, can be circumvented not only by introducing surface irregularities but also by chemisorption. On the basis of chemisorbed atoms Stecher and Williams (1966) have calculated the formation rates for some diatomic molecules and radicals on the surface of graphite and dirty ice grains. The important mechanism of interstellar molecule formation is the capture by an incident atom of an atom chemically bound to the grain, to form a molecule by an exothermic chemical exchange reaction. The reactions involved are then surface reactions and chemical exchange reactions. Empty valences of the grains become occupied by atoms of the interstellar gas, which are then chemically bound to the surface of the grain. Further collisions by interstellar atoms with these surface complexes may lead to molecular rearrangement and in the case of exothermal reactions to a release of molecules to the interstellar medium. A general reaction of this type may be written:

$$\text{grain} \cdot X + Y \rightarrow \text{grain} + XY \tag{39}$$

Two different forms of grains have been specifically discussed: in the first case formation on graphite is considered, the grains of which may be represented by a hexagon ring.

$$\tag{40}$$

The C atom is a member of the graphite grain and serves to chemisorb the X atom until molecule formation has taken place. The theory used in the calculations of formation rates on grains is collision theory (*e.g.* Polanyi, 1962). As always assumed it is dependent on an activation energy $A$, which may be regarded as the energy the incoming atom requires to overcome the repulsion of the chemically bound molecule. The reaction rate is then proportional to

$$n_Y \cdot n_{\text{molecule}}\, \sigma^2\, T^{1/2} \exp\left(-A/kT\right) \tag{41a}$$

where $n_Y$ and $n_{molecule}$ are the number densities of the incident atoms and the molecule. If this equation is applied to the special cases considered in Eqs. (39) and (40) the molecular production rates can be written in a simplified form

$$R_{gr} = b \, T^{1/2} \, e^{-a/T} \cdot p^2 \cdot f(T) \cdot \xi(T) \qquad (41b)$$

where $a$ and $b$ are constants, $p$ is the pressure in units of $n_H = 10 \ \text{cm}^{-3}$, $f(T)$ is the fraction of atoms available on the grain surface for exchange and $\xi(T)$ is a function allowing for the increase in grain cross-section for a negatively charged grain approached by a positive ion ($\xi = 1$ for neutral atoms). Various cases have been considered and are listed in Table 12.

Table 12. Formation of molecules on graphite grains (Stecher and Williams, 1966)

| X \ Y | H | $C^+$ | O | N |
|---|---|---|---|---|
| H | $H_2$ | CH[1] | OH | NH |
| $C^+$ | | | CO[1] | CN[1] |
| N | | | | $N_2$ |

[1]) The excess charge is taken up by the grain.

In the second case molecule formation on dirty ice grains is considered and found to be rather limited in that it produces only $H_2$ and CO since all other reactions are endothermic. However these two molecules seem to be most widely distributed.

The predicted equilibrium densities calculated under the assumption that the destruction processes are mainly chemical exchange reactions and photo-dissociation are not in very good agreement with the observed densities. This may be due to the fact that only graphite and dirty ice grains were considered and that other grain surfaces may have to be considered as well. Furthermore the possibility of polyatomic molecule formation during these reactions cannot be excluded and may therefore if properly taken into account alter the derived results.

In a more recent paper by Watson and Salpeter (1972a) the formation of interstellar molecules on the surface of grains is also discussed.

It may be concluded from their calculations that strong chemisorption of atoms and radicals prevents not only molecule formation but ejection or desorption from the grain surface as well. However, on top of the chemisorbed layer physical adsorption is possible and the adsorbed particles can move over the whole grain surface either by thermal motion or by quantum mechanical barrier penetration. Since hydrogen has the highest surface abundance the most common reactions will be the formation of diatomic radicals like CH, NH and OH.

Assuming that a large fraction of these radicals remains at the surface, they can further react to form saturated molecules like $CH_4$, $NH_3$ or $H_2O$. Reactions of these molecules with radicals, which have some excitation energy, either as a consequence of their formation or from the photodissociation of saturated molecules, can then lead to more complex organic molecules. The branching ratio, which determines the chemical composition of the products thus formed, depends on the surface abundance of atoms and radicals and any possible ejection mechanisms which may interrupt the reaction sequence.

The ejection mechanisms considered by Watson and Salpeter are mainly photodesorption and ejection during molecule formation, the former being the most important process. Other mechanisms involving interaction with cosmic rays or with IR photons are of minor importance. Since ejection by UV photons is highly unlikely in very dense clouds, condensation of the interstellar gas and grain growth in the Watson-Salpeter-model would proceed without limit, which of course is in contradiction to observations.

In a second paper Watson and Salpeter (1972b) calculate the abundance of interstellar molecules as a function of the interstellar radiation field. For unshielded regions Habing's (1968) calculated values for the interstellar UV radiation density are used (see Table 3). The connection between the gas density and the shielded UV light is established by introducing the parameter

$$\xi = \frac{n}{100} \ \exp \ \{2.5 \ \tau_\nu\} \ ,$$ $n$ being the gas density. For photons with energies above 11.3 eV the shielding factor increases rapidly because of photoabsorption by carbon.

Relative abundances are calculated by assuming that every atom or molecule with an atomic weight $\geqslant 12$ hitting an adsorbed atom or radical forms some molecule. This leads to a rate of formation

$$R_{gr} = (\frac{n}{100}) \cdot 10^{-15} \ sec^{-1} \tag{42}$$

In clouds with $\xi < 10$ the combined abundances for molecules containing C, N and O relative to atomic C, N and O are $\sim 10^{-4}\xi$. For moderate and heavy shielding with $\xi > 30$ Watson and Salpeter predict

$$\frac{[CH]}{[C]} \simeq \frac{[NH]}{[N]} \simeq \frac{[OH]}{[O]} \simeq \frac{[SH]}{[S]} \simeq \frac{3 \cdot 10^{-4}}{(1 + 250 \ \xi^{-1})} \tag{43}$$

For the other molecules and for $\xi \geqslant 200$ the following ratios are obtained

$$\frac{[CO]}{[C]} \sim 10^{-4} \, \xi^2 \tag{44}$$

$$\frac{[CN]}{[C]} \sim 5 \, \frac{[CS]}{[C]} \sim \frac{[N_2]}{[N]} \sim 3 \, \frac{[NO]}{[N]} \sim 10^{-5} \, \xi \tag{45}$$

The high predicted abundance of CO is in agreement with the observational results. The fact that NO has not been found in interstellar space may be due to exchange reactions like $N + NO \rightarrow N_2 + O$ with appreciable rate constants. The observed ratio [HCN] / [CN] is larger than the predicted value and suggests either that HCN is produced directly on the grain or that CN destruction is faster than estimated.

The result of these calculations shows that even in "normal" HI clouds with $\xi \sim 1$ and no strong extinction appreciable molecular abundances can exist. The predicted values are:

[OH] / [H] $\sim 2 \times 10^{-8}$; [CH] / [H] $\sim 10^{-9}$; [H$_2$O] / [H] $\sim 10^{-9}$;
[CO] / [H] $\sim 4 \times 10^{-8}$; [CN] / [H] $\sim 4 \times 10^{-9}$

Among other mechanisms which have been proposed for molecule production in interstellar space three are worth discussing here briefly. According to Sagan (1972) the organic material produced in the early history of the solar system can be making contributions to interstellar organic chemistry because of the dissipation of organic molecules during star formation and the subsequent loss of comets to the interstellar medium. Hence the observed small molecules may be degradation products of larger organic molecules.

The storage of highly reactive radicals in ice grains has been discussed by Greenberg (1971). On local heating of these grains by cosmic rays or UV photons, the radicals may start a chain reaction which leads to the explosion of the whole grain. During the explosion complex organic molecules would be formed and ejected into interstellar space.

The synthesis of organic molecules in Fischer-Tropsch reactions in early solar nebulae has been suggested by Anders et al. (1971).

The difficulty with the mechanisms suggested by Sagan and by Anders is that molecules are usually found in dense clouds, where they are shielded from the dissociating radiation. The lifetimes of the relevant molecules in unshielded regions are too short to let them travel from any place outside a cloud into the denser regions where they are observed.

### D. Laboratory Experiments

At the moment very little information from laboratory-experiments is available which is directly applicable to interstellar chemistry. Because of their temperature and pressure dependence, such well-known catalytic processes as Fischer-

Tropsch etc. can under interstellar conditions hardly lead to formation rates which would explain the observed abundances. Especially in the formation of the more complex molecules the interstellar radiation field and the dust grains seem to play an important role. In experiments simulating the Martian atmosphere Hubbard $et\ al.$ (1971) found that $^{14}CO_2$ and $^{14}C$-organic compounds are formed when a dilute mixture of $^{14}CO$ and water vapour in $^{12}CO_2$ or $N_2$ is irradiated with UV light in the presence of soil or pulverized vycor. Among other organic compounds formaldehyde and acetaldehyde could be identified. The formation of the organic compounds occurs over a broad spectral range below 3000 Å.

In order to investigate the primary processes responsible for "photocatalytic" reactions Moesta and Breuer (1968) performed experiments in which simple gases, some examples of which are given later, were adsorbed on clean metal surfaces and subsequently irradiated by UV light. The different metals used in these experiments certainly do not resemble the chemical and physical composition of interstellar dust grains, but they at least offer well defined surfaces and thus avoid introducing errors due to impurities which may lead to a misunderstanding of the primary processes.

The main results of these experiments may be summarized in two points:
(1) as soon as adsorption occurs the absorption frequencies of the electronic transitions of the primary molecules are shifted to longer wavelengths. This fact is then used to irradiate the adsorbed molecules with light corresponding to the shifted wavelength.
(2) Irradiation with wavelengths $\lambda > 2000$ Å leads to the formation of reactive atoms or radicals on the surface which in turn can react with other adsorbed species to form rather complex molecules.

Out of a large number of experiments three selected examples are presented here. In a first experiment pure CO was adsorbed on a clean tungsten surface and was irradiated by light of the wavelength $\lambda$ 2537 Å. The results of the mass spectroscopic study of the observed reaction products are summarized in Table 13.

Table 13. Reaction scheme and mass spectrum of CO on tungsten-surface (H. D. Breuer and H. Moesta, 1971)

| | | |
|---|---|---|
| CO | $+ h\nu$ | $\rightarrow CO^*$ |
| $CO^*$ | | $\rightarrow C + O$ |
| C | $+ CO$ | $\rightarrow C_2O$ |
| O | $+ CO$ | $\rightarrow CO_2$ |
| O | $+ C_2O$ | $\rightarrow C_2O_2$ |
| C | $+ C_2O_2$ | $\rightarrow C_3O_2$ |

(The first two rows are bracketed together as "primary step")

It is assumed that in the primary reaction electronically excited $CO^*$ is formed and decays into C and O atoms which then react with CO or the already formed molecular species to give the reaction products indicated in the last four lines of Table 13. These end products can be monitored by their mass spectrum. Because of its low adsorption energy $CO_2$ is desorbed almost completely while the other products are only partially desorbed during the formation reaction.

In a second experiment a mixture of adsorbed $CH_4$ and CO in equal parts was irradiated by radiation of $\lambda$ 2537 Å. In this reaction the formation of $H_2CO$ could be identified by its mass spectrum. This was clearly substantiated by an additional experiment, in which the isotopic mass of $D_2CO$ was detected when $CH_4$ was replaced by $CD_4$ as primary reaction partner.

As the number of primary reaction partners is increased, the observed mass spectrum becomes rather complex and ambiguities occur in identification. As a last example, we mention the photoproducts obtained by subjecting co-adsorbed $H_2$, CO and $N_2$ (in approximately equal parts) on tungsten surface to UV radiation of $\lambda$ = 2537 Å.

Table 14. Mass spectrum obtained by irradiating $H_2$, CO and $N_2$ on tungsten-surface

| $m/e$ | Molecule |
| --- | --- |
| 26 | CN |
| 27 | HCN |
| 29 | $N_2H$ |
| 30 | NO, $N_2H_2$ |
| 38 | $C_2N$ |
| 39 | $C_2HN$ |
| 40 | $C_2H_2N$ |
| 41 | $CH_3CN$ |
| 43 | HNCO |
| 44 | $CO_2$, $N_2O$ |
| 46 | $NO_2$ |
| 47 | $HCONH_2$ |
| 51 | $HC_3N$ |

In some cases the cross sections for molecule formation and desorption from the surface could be estimated from the measured intensity of the mass peaks and the known photon flux at the surface. The corresponding values are listed in Table 15.

The fact that most of the observed photocatalytic reactions occur in the wavelength region between 2000 Å and 3000 Å is important for the application of these reactions to interstellar chemistry, since photons in this wavelength range can penetrate even dense clouds. The values for the cross-sections indicate that photocatalytic reactions may be a very effective mechanism in producing interstellar molecules.

Table 15. Cross sections in $cm^2$ for
formation and desorption at $\lambda$ =
2537 Å (H. D. Breuer, 1971).

| | | |
|---|---|---|
| $\sigma$(CN) | = | $1.5 \times 10^{-18}$ |
| $\sigma$(HCN) | = | $5 \times 10^{-19}$ |
| $\sigma$(H$_2$CO) | = | $1 \times 10^{-16}$ |
| $\sigma$(CH$_3$CN) | = | $1.5 \times 10^{-18}$ |
| $\sigma$(HNCO) | = | $1 \times 10^{-18}$ |
| $\sigma$(HCONH$_2$) | = | $1 \times 10^{-18}$ |
| $\sigma$(HC$_3$N) | = | $5 \times 10^{-19}$ |

Assuming a grain diameter of $0.1\mu$, a cloud corresponding to $\xi \sim 1000$ in the Watson-Salpeter model, and a mean cross-section of $10^{-18}$ $cm^2$ the mean abundance of complex molecules has been calculated as an equilibrium between production by photochemical surface reactions and destruction (see IV. E) by absorption of UV photons (Breuer, 1971). For a hydrogen density of $n_H$ = $10^2$ $cm^{-3}$ an average molecule density of $4.5 \times 10^{-6}$ $cm^{-3}$ was obtained, a value which is in agreement with the observed densities for at least some regions in space.

## E. Lifetime and Destruction of Interstellar Molecules

Photodissociation and exchange reactions are the primary destruction process for interstellar molecules. If photodecomposition is considered to be the only destruction mechanism, then the lifetime of interstellar molecules depends upon three factors: the absorption cross section, the quantum yield of dissociation and the interstellar radiation field. A quantitative discussion of this destruction mechanism has been given by Stief (1971) for two diatomic and eight polyatomic molecules.

During photodecomposition of the molecules $H_2CO$, $NH_3$, $H_2O$ and $CH_4$ the dominant primary processes encountered are the formation of atomic and molecular hydrogen; the relative importance of the two decomposition channels is strongly dependent on the wavelength. Available laboratory data indicate that these and all other polyatomic molecules are dissociated in the wavelength region longer than 912 Å. CO however, has an exceptionally high decomposition threshold ($\lambda$ = 1115 Å) and is thus the most stable of all known interstellar molecules. Unfortunately little laboratory data are available for the range between the decomposition threshold of CO and 912 Å. The decomposition of OCS for example into CO and S has a quantum yield of 0.9 at $\lambda$ = 2537 Å and $\lambda$ = 2288 Å. Formation of atomic oxygen and CS becomes energetically possible at wavelengths shorter than $\lambda$ = 1735 Å. Included in Stief's calculations are molecules like NO, $C_2H_2$, $CH_4$ and $C_6H_6$. Unfortunately, of these molecules, only NO has an observable radio spectrum and despite an extensive interstellar search NO has not been detected.

The dissociation probability of the molecules is given by

$$p = \frac{u_\lambda \Phi}{h} \int \theta_\lambda \, \sigma_\lambda \cdot \lambda d\lambda \qquad (46)$$

where $u_\lambda$ is the radiation density, $\Phi$ is the quantum yield for dissociation, $h$ is Planck's constant, $\theta_\lambda$ is the transmissivity of a cloud which has been evaluated from the extinction in the visual, $A_v$, and $\sigma_\lambda$ is the absorption cross section. The result of these calculations is shown in Figs. 21a and b. It is clear from these figures that the lifetimes of all molecules with exception of carbon monoxide are $\leqslant 100$ years in unshielded regions. The total distance traveled by a molecule in its lifetime of 100 years is in the order of 0.001 pc. Since intercloud distances are appreciably greater than this value it is evident that a polyatomic molecule cannot travel appreciably more than 0.001 pc in unobscured regions without being photodissociated. As a consequence, such molecules can only be formed or released in the cloud where they are observed.

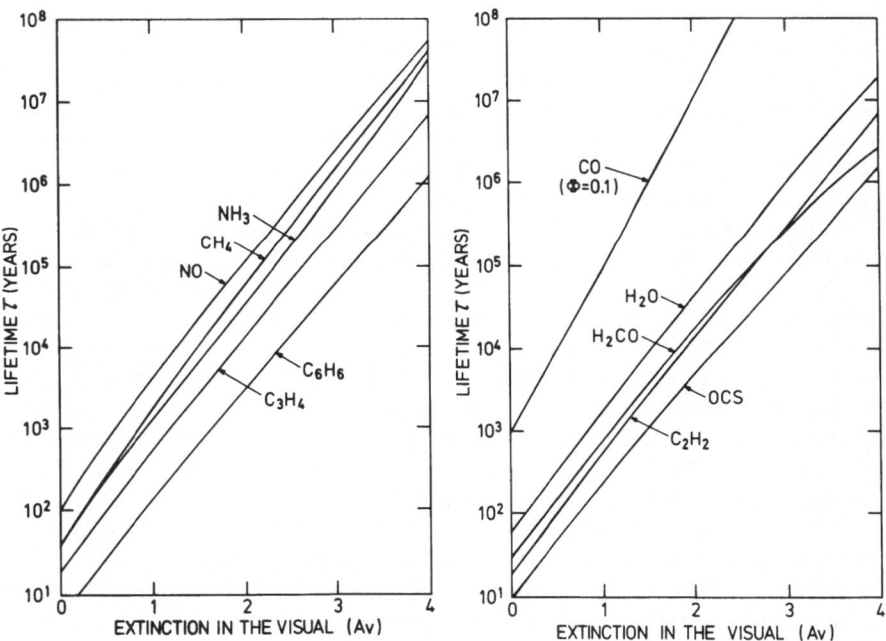

Fig. 21a and b. Lifetime of interstellar molecules as a function of the thickness of the shielding dust layer, expressed in magnitudes of visual extinction (after Stief, 1971)

In addition to the calculated lifetime of photodissociation a contribution from photoionization might be expected. For the ten molecules considered, photons with $\lambda > 912$ Å are capable of ionizing all molecules except CO. However, ionization will in general be less important than dissociation since it occurs not only over a narrower wavelength range but the ionization efficiencies are also less than unity for polyatomic molecules.

The destruction of interstellar molecules by interaction with soft X-rays and subcosmic rays is even less important, because of the general decrease of the decomposition cross section with increasing energy of the collision partner.

## F. Summary

Since the detection of interstellar molecules, several mechanisms have been proposed to explain their presence in the interstellar medium. Formation, relaxation and destruction mechanisms are important processes in the life cycle of an interstellar molecule. The formation processes discussed involve gas-phase- and surface reactions, the latter taking place upon interstellar dust grains. Several types of gas-phase reactions are proposed such as radiative recombination, ion-neutral, neutral-neutral, chemical-exchange and recombination reactions. The chemistry of these interstellar clouds depends upon, whether a low to medium or high density region is discussed. In the latter region molecules dominate, whereas atoms are the main constituent of low density regions and photodissociation becomes important. Rate constants are either experimentally determined or theoretically calculated. Their precise knowledge and the assurance that all relevant reactions are included in the treatment pose serious difficulties to all gas-phase calculations. Furthermore, all gas-phase models are sensitive to the presence of interstellar dust, which shields the molecules from the destructive interstellar UV radiation field.

It is almost certain that the formation of $H_2$ from atomic hydrogen occurs on interstellar grain surfaces. Less understood, however, is the formation of other molecular species on grain surfaces, since some of the relevant physical processes on the surfaces of interstellar dust grains are highly uncertain. These processes include: condensation of the gas onto grains, sticking probability, surface mobility, reaction rates and others. For example, the ejection of adsorbed molecules from the grain surface may be due to interaction with UV photons or by converting the heat of formation into kinetic energy sufficient for desorption. Interaction with the interstellar radiation field may also lead to molecule formation through photocatalytic reactions.

The time elapsed between hitting the grain surface and molecule formation is of the order of seconds. The lifetime of the molecule in free space after ejection from the surface depends upon the region where it is released. In a cloud with a visual extinction of $A_v = 3$ corresponding to a hydrogen column density of $N_H = 5 \times 10^{21}$ cm$^{-2}$ the lifetime of a typical molecule will be in the order of $10^{13}$ sec. During this time the probability for photodissociation is unity. The dissociation products are then available for recycling.

Although it is presently not possible to distinguish clearly between the different formation mechanisms, it is felt that more observational data will furnish the material necessary for that. There are two areas of immediate help which play a very important role in our understanding of interstellar chemistry: Isotopic abundance determinations, notably deuterium to hydrogen ratios, and the identification of the unknown interstellar lines by astronomical, laboratory or theoretical means.

## V. Appendix

This compilation covers all molecular lines detected by radioastronomical techniques in the interstellar medium. The listing includes all lines reported up to August 1973. The spectroscopic notation corresponds to the one used in Table 6. The electronic ground state is indicated only in cases where it is different from a $^1\Sigma$ state. Each entry under the six sources listed means that this particular transition has been observed in that source.

## Appendix

| Molecule | Transition | Frequency MHz | E/A | Sgr B2 | Sgr A | Ori A | W51 | W3 | DR 21 |
|---|---|---|---|---|---|---|---|---|---|
| OH $^2\Pi_{3/2}$ | $J = 3/2, F = 1 - 2$ | 1612.231 | A/E | * | * | * | * | * | * |
| | $1 - 1$ | 1665.401 | A/E | * | * | * | * | * | * |
| | $2 - 2$ | 1667.358 | A/E | * | * | * | * | * | * |
| | $2 - 1$ | 1720.533 | A/E | * | * | * | * | * | |
| | $J = 5/2 \quad 2 - 2$ | 6030.739 | E | | | | | * | |
| | $3 - 3$ | 6035.085 | E | * | | | | * | |
| | $J = 7/2 \quad 4 - 4$ | 13441.371 | E | | | | | * | * |
| OH $^2\Pi_{1/2}$ | $J = 1/2, F = 0 - 1$ | 4660.242 | E | * | * | | | | |
| | $1 - 1$ | 4750.656 | E | * | | | | | |
| | $1 - 0$ | 4765.562 | E | * | | * | | | |
| $^{18}$OH $^2\Pi_{3/2}$ | $J = 3/2, F = 1 - 1$ | 1637.53 | A | * | * | | | * | |
| | $2 - 2$ | 1639.48 | A | * | * | | | | |
| SiO | $J = 2 - 1$ | 86847. | E | * | * | | | | |
| | $3 - 2$ | 130268.4 | E | * | | | | | |
| SO $^3\Sigma^-$ | $J,N = 3,2 - 2,1$ | 99299.85 | E | * | * | * | * | * | * |
| | $4,3 - 3,2$ | 138178.60 | E | * | * | * | * | * | * |
| $H_2O$ | $6_{16} - 5_{23}$ | 22235.07985 | ME | * | * | * | * | * | * |
| $H_2S$ | $1_{10} - 1_{01}$ | 168762.762 | E | * | * | * | * | * | * |
| $NH_3$ | $J,K = 1,1$ | 23694.48 | E | * | * | * | * | * | * |
| | $2,2$ | 23722.71 | E | * | * | * | * | | |
| | $3,3$ | 23870.11 | E | * | * | * | | | |
| | $4,4$ | 24139.39 | E | * | | | | | |

Appendix (continued)

| Molecule | Transition | Frequency MHz | E/A | Sgr B2 | Sgr A | Ori A | W51 | W3 | DR21 |
|---|---|---|---|---|---|---|---|---|---|
| | 6,6 | 25056.04 | E | * | | | | | |
| | 2,1 | 23098.78 | E | * | | | | | |
| | 3,2 | 22834.10 | E | * | | | | | |
| CN $^2\Sigma^-$ | $N = 1 - 0, J = 3/2 - 1/2$ 113491. | E | * | * | * | | | | |
| CO | 1 − 0 | 115271.204 | E | * | * | * | * | | * |
| | 2 − 1 | 230537.974 | E | | * | * | * | * | * |
| $^{13}$CO | 1 − 0 | 110201.370 | E | * | * | | | * | |
| C$^{18}$O | 1 − 0 | 109782.182 | E | * | | * | * | | |
| C$^{17}$O | 1 − 0 | 112359.276 | E | * | | | | | |
| CS | 1 − 0 | 48991.000 | E | * | * | * | | | * |
| | 2 − 1 | 97981.007 | E | * | | | | | |
| | 3 − 2 | 146969.039 | E | | * | * | | | * |
| $^{13}$CS | 1 − 0 | 46247.47 | E | * | | | | | |
| | 2 − 1 | 92493.95 | E | * | | | | | |
| C$^{34}$S | 1 − 0 | 48206.948 | E | * | | | | | |
| | 2 − 1 | 96412.953 | E | * | | | | | |
| OCS | 7 − 6 | 85139.11 | E | * | | | | | |
| | 8 − 7 | 97301.22 | E | * | | | | | |
| | 9 − 8 | 109463.063 | E | * | | | | | |
| | 12 − 11 | 145946.821 | E | * | | | | | |

Appendix (continued)

| Molecule | Transition | Frequency MHz | E/A | Sgr B2 | Sgr A | Ori A | W51 | W3 | DR21 |
|---|---|---|---|---|---|---|---|---|---|
| HCN | $1-0, F=1-1$ | 88630.416 ⎫ | E | * | * | * | * | * | * |
| | $2-1$ | 88631.847 ⎬ | | | | | | | |
| | $0-1$ | 88633.936 ⎭ | | | | | | | |
| H$^{13}$CN | $1-0, F=1-1$ | 86338.75 ⎫ | E | | * | * | | | |
| | $2-1$ | 86340.05 ⎬ | | | | | | | |
| | $0-1$ | 86342.16 ⎭ | | | | | | | |
| HC$^{15}$N | $2-1$ | 172677.1 | E | | | * | | | * |
| | $2-1$ | 172108.1 | E | | * | * | | | * |
| DCN | $1-0$ | 72414.7 | E | | * | * | | | |
| | $2-1$ | 144828.0 | E | | * | * | | | |
| H$_2$CO | $1_{10}-1_{11}$ | 4829.660 | A | * | * | * | * | * | * |
| | $2_{11}-2_{12}$ | 14488.650 | A | * | * | * | * | | |
| | $3_{12}-3_{13}$ | 28974.850 | A | * | | | | | |
| | $5_{14}-5_{15}$ | 72409.099 | E | | | * | | | |
| | $1_{01}-0_{00}$ | 72837.974 | E | | | * | | | |
| | $2_{02}-1_{01}$ | 145602.971 | E | | | * | | | |
| | $2_{12}-1_{11}$ | 140839.529 | E | | * | * | * | * | |
| | $2_{11}-1_{10}$ | 150498.359 | E | | | * | | | |
| H$_2^{13}$CO | $1_{10}-1_{11}$ | 4593.089 | A | * | * | * | * | | |
| H$_2$C$^{18}$O | $1_{10}-1_{11}$ | 4388.797 | A | * | | * | | | |

Appendix (continued)

| Molecule | Transition | Frequency MHz | E/A | Sgr B2 | Sgr A | Ori A | W51 | W3 | DR21 |
|---|---|---|---|---|---|---|---|---|---|
| $H_2CS$ | $1_{10} - 1_{11}$ | 1046.48 | A | | | | * | | |
| | $2_{11} - 2_{12}$ | 3139.38 | A | * | | | | | |
| $HNCO$ | $1_{10} - 0_{00}$ | 21981.5 | E | * | | | | | |
| | $4_{04} - 3_{03}$ | 87925.24 | E | * | | | | | |
| | $4_{13} - 3_{12}$ | 88239.05 | E | * | | | | | |
| | $5_{05} - 4_{04}$ | 109905.85 | E | * | | | | | |
| $HCOOH$ | $1_{10} - 1_{11}$ | 1638.805 (?) | E | * | | | | | |
| $HCCCN$ | $1-0, F=1-1$ | 9097.036 | E | * | | | | | |
| | $2-1$ | 9098.332 | E | * | | | | | |
| | $0-1$ | 9100.279 | E | * | | | | | |
| | $8-7$ | 72783.8 | E | * | | | | | |
| | $9-8$ | 81881.5 | E | * | | | | | |
| | $10-9$ | 90979.0 | E | * | * | * | | | |
| | $11-10$ | 100076.4 | E | * | | | * | | |
| $H_2CNH$ | $1_{10} - 1_{11}, F = 0-1$ | 5288.980 | | * | | | | | |
| | $1-0$ | 5289.786 | | * | | | | | |
| | $2-2$ | 5289.786 | E | * | | | | | |
| | $2-1$ | 5290.726 | | * | | | | | |
| | $1-2$ | 5290.726 | | * | | | | | |
| | $1-1$ | 5291.646 | | * | | | | | |

Appendix (continued)

| Molecule | Transition | Frequency MHz | E/A | Sgr B2 | Sgr A | Ori A | W51 | W3 | DR21 |
|---|---|---|---|---|---|---|---|---|---|
| CH$_3$OH | 1,1 – 1,1 A | 834.267 | E | * | * | | | | |
| | 3,1 – 3,1 A | 5005.37 | E | * | | | | | |
| | 1,0 – 0,0 A | 48372.60 | E | * | | | | | |
| | 1,0 – 0,0 E | 48377.09 | E | * | | | | | * |
| | 3,0 – 2,0 E | 145093.75 | E | * | * | * | | | * |
| | 3,1 – 2,1 E$_2$ | 145097.47 | E | * | * | * | | | * |
| | 3,0 – 2,0 A | 145103.23 | E | * | * | * | | | * |
| | 3,2 – 2,2 A | 145124.41 | E | | | * | | | * |
| | 3,2 – 2,2 E$_1$ | 145126.37 | E | | | * | | | * |
| | 3,2 – 2,2 E$_2$ | 145126.37 | E | | | * | | | * |
| | 3,1 – 2,1 E$_1$ | 145131.88 | E | | | * | | | * |
| | 3,2 – 2,2 A | 145133.46 | E | | | * | | | |
| | 4,2 – 4,1 E$_1$ | 24933.47 | E | | | * | | | |
| | 2,2 – 2,1 E$_1$ | 24934.38 | E | | | * | | | |
| | 5,2 – 5,1 E$_1$ | 24959.08 | E | | | * | | | |
| | 6,2 – 6,1 E$_1$ | 25018.14 | E | | | * | | | |
| | 7,2 – 7,1 E$_1$ | 25124.88 | E | | | | | | |
| | 8,2 – 8,1 E$_1$ | 25294.41 | E | | | * | | | |
| | 5,1 – 4,0 E$_2$ | 84521.21 | E | * | | | | | |
| | 10,1 – 10,0 E$_1$ | 169335.34 | | * | | | | | |
| CH$_3$CN | 6,0 – 5,0 | 110383.5 | E | * | | | | | |
| | 6,1 – 5,1 | 110381.4 | E | | | | | | |
| | 6,2 – 5,2 | 110375.0 | E | * | | | | | |
| | 6,3 – 5,3 | 110364.5 | E | * | | | | | |

Appendix (continued)

| Molecule | Transition | Frequency MHz | E/A | Sgr B2 | Sgr A | Ori A | W51 | W3 | DR21 |
|---|---|---|---|---|---|---|---|---|---|
| CH₃CN | $6,4 - 5,4$ | 110349.7 | E | * | * | | | | |
| | $6,5 - 5,5$ | 110330.7 | E | * | * | | | | |
| HCONH₂ | $1_{10} - 1_{11}, F = 1 - 1$ | 1538.135 | E | * | * | | | | |
| | $1 - 2$ | 1538.693 | E | * | * | | | | |
| | $2 - 1$ | 1539.295 | E | * | | | | | |
| | $1 - 0$ | 1539.570 | E | * | | | | | |
| | $2 - 2$ | 1539.851 | E | * | * | | | | |
| | $0 - 1$ | 1541.018 | E | * | * | | | | |
| | $2_{11} - 2_{12}, F = 2 - 2$ | 4617.14 | E | * | | | | | |
| | $3 - 3$ | 4619.00 | E | * | | | | | |
| | $1 - 1$ | 4620.01 | E | * | | | | | |
| CH₃CHO | $1_{10} - 1_{11}$ | 1065.075 | E | * | * | | | | |
| | $2_{11} - 2_{12}$ | 3195.167 | E | * | | | | | |
| CH₃CCH | $5,0 - 4,0$ | 85457.29 | E | * | | | | | |
| | $5,1 - 4,1$ | 85455.67 | E | * | | | | | |
| | $5,2 - 4,2$ | 85450.78 | E | * | | | | | |
| | $5,3 - 4,3$ | 85442.61 | E | * | | | | | |
| U81.5 | | 81541. | E | * | | | | | |
| U85.4 | | 85435. | E | * | | | | | |
| U89.2 | | 89190. | E | | * | * | * | * | * |
| U90.7 | | 90663.9 | E | | | * | | | |
| U144.9 | | 144858 | E | * | | | | | |

79

G. Winnewisser, P. G. Mezger and H.-D. Breuer

## VI. References

Adams, W. S.: Astrophys. J. *93*, 11 (1941).
Anders, E.: Proc. of Symposium on Interstellar Molecules, Charlottesville, Va. 1971.
Bates, D. R.: Mon. Not. R. Astr. Soc. *111*, 303 (1951).
Breuer, H. D.: Habilitationsschrift. Saarbrücken: Universität des Saarlandes 1971.
Breuer, H. D., Moesta, H.: Highlights of astronomy (ed. C. de Jager) 1970.
Breuer, H. D.: Proc. of Symposium on Interstellar Molecules, Charlottesville, Va. 1971.
Buhl, D., Snyder, L. E., Edrich, J. S.: Astrophys. J. *177*, L 625 (1972).
Carruthers, G. R.: Astrophys. J. *161*, L81 (1970).
Cesarsky, D. A., Moffet, A. T., Pasachoff, J. M.: Astrophys. J. *180*, L 1 (1973).
Cheung, A. C., Rank, D. M., Townes, C. H., Thornton, D. D., Welch, W. J.: Phys. Rev. Lett. *21*, 1701 (1968).
Cheung, A. C., Rank, D. M., Townes, C. H., Welch, W. J.: Nature *221*, 917 (1969).
Dorney, A. J., Watson, J. K. G.: J. Mol. Spectrosc. *42*, 135 (1972).
Field, G. B. in: Proc. of the 16th Liege Symposium, p. 29. Inst. d'Astrophysique de l'Universite de Liege 1970.
Giaconni, R., Gursky, H., van Speybroeck, L. P.: Ann. Rev. Astron. Astrophys. *6*, 373 (1968).
Gardner, F. F., Ribes, J. C., Cooper, B. F. C.: Astrophys. Letters *9*, 181 (1971).
Gordy, W., Cook, R. L.: Microwave molecular spectra, in: Chemical applications of spectroscopy, Vol. IX, Part II. New York: Wiley Interscience 1970.
Greenberg, M.: Private communication (1971).
Habing, H. J.: Bull Astr. Insts. Neth. *19*, 421 (1968).
Hills, R., Janssen, M. A., Thornton, D. D., Welch, W. J.: Astrophys. J. *175*, L59 (1972).
Heiles, C.: Ann. Rev. Astron. Astrophys. *9*, 293 (1971).
Herbig, G. H.: Z. f. Astrophysik *68*, 243 (1968).
Herzberg, G., Johns, J. W. C.: Astrophys. J. *158*, 399 (1969).
Hills, R., Janssen, M. A., Thornton, D. D., Welch, W. J.: Astrophys. J. *175*, L 59 (1972).
Hollenbach, D. J., Salpeter, E. E.: Astrophys. J. *163*, 155 (1971).
Hollenbach, D. J., Werner, M. W., Salpeter, E. E.: Astrophys. J. *163*, 165 (1971).
Hubbard, J. S., Hardy, J. P., Horowitz, N. H.: Proc. Natl. Acad. Sci. *68*, 574 (1971).
Jefferts, K. B., Penzias, A. A., Wilson, R. W.: Astrophys. J. *179*, L 57 (1973).
Jong de, T.: Astron. Astrophys. *20*, 263 (1972).
Kley, D., Washida, N., Becker, K. H., Groth, W.: Z. f. Phys. Chem. N. F. *82*, 109 (1972).
Lynds, B. T.: Astrophys. J. Suppl. Ser. *7*, 1 (1962).
Litvak, M. M.: Astrophys. J. *160*, L 133 (1970).
Lucia de, F. C., Helminger, P., Cook, R. L., Gordy, W.: Phys. Rev. *5A*, 487 (1972).
McKellar, A.: Publ. Astr. Soc. Pacific *52*, 187 (1940).
Mezger, P. G., in: Highlights of Astronomy (ed. C. de Jager), Vol. 2, p. 366 (1971).
Mezger, P. G., in: Proc. 2nd Adv. Course in Astr. and Astrophys. Interstellar Matter: An Observes View. Geneva Observatory 1972.
Mezger, P. G., Robinson, B. J.: Nature *220*, 1107 (1968).
Moesta, H., Breuer, H. D.: Naturwissenschaften *55*, 650 (1968).
Oka, T.: J. Chem. Phys. *49*, 3135 (1968). – Daly, P. W., Oka, T.: J. Chem. Phys. *53*, 3272 (1970).
Oka, T.: Astrophys. J. *160*, L 69 (1970).
Oka, T., Shimizu, F. O., Shimizu, T., Watson, J. K. G.: Astrophys. J. *165*, L 15 (1971).
Oka, T.: Proc. Symp. on Interstellar Molecules, Charlottesville, Va. 1971.
Palmer, P., Zuckerman, B., Buhl, D., Snyder, L. E.: Astrophys. J. *156*, L 147 (1969).
Penzias, A. A., Wilson, R. W.: Astrophys. J. *142*, 419 (1965).
Penzias, A. A., Jefferts, K. B., Wilson, R. W.: Astrophys. J. *165*, L 229 (1971).
Penzias, A. A.: Private communication (1973).
Polanyi, J. C.: Chemical processes, in: Atomic and molecular processes (ed. D. R. Bates). New York: Academic Press 1962.
Poletto, G., Rigutti, M.: Nuovo Cimento *39*, 515 (1965).

Rank, D. M., Townes, C. H., Welch, W. J.: Science *174*, 1083 (1971).
Rogerson, J. B., York, D. G.: Comm. Int. Astr. Union, Sept. (1973).
Sagan, C.: Nature *238*, 77 (1972).
Savage, B. D., Jenkins, E. B.: Astrophys. J. *172*, 491 (1972).
Scoville, N. Z.: Unpubl. thesis. New York: Columbia University 1972.
Simonson III, S. Ch.: Astron. Astrophys. *9*, 163 (1970).
Smith, A. M., Stecher, T. P.: Astrophys. J. *164*, L 43 (1971).
Snyder, L. E.: MTP Review of Chemistry, Phys. Chem. Series 1, Vol. 3, Spectroscopy (ed. D. A. Ramsay). Oxford: Medical and Technical Publ. Co. Ltd. and London: Butterworth Co., Ltd. 1972.
Snyder, L. E., Buhl, D.: Astrophys. J. (to be published) (1971).
Snyder, L. E., Buhl, D., Zuckerman, B., Palmer, P.: Phys. Rev. Lett. *22*, 679 (1969).
Solomon, P. M., Jefferts, K. B., Penzias, A. A., Wilson, R. W.: Astrophys. J. *168*, L 107 (1971).
Solomon, P. M., Klemperer, W.: Astrophys. J. *178*, 389 (1971).
Solomon, P. M., Thaddeus, P.: (to be published) (1971).
Spitzer, L., Drake, J. F., Jenkins, E. B., Morton, D. C., Rogerson, J. B., York, D. G.: Astrophys. J. *181*, L 116 (1973).
Stecher, T. P., Williams, D. A.: Astrophys. J. *146*, 88 (1966).
Stecher, T. P., Williams, D. A.: Astrophys. J. *177*, L 141 (1972).
Stief, L. J.: Proc. Symposium on Interstellar Molecules, Charlottesville, Va. 1971.
Stief, L. J., Donn, B., Glicker, S., Gentien, E. P., Mentall, J. E.: Astrophys. J. *171*, 21 (1972).
Strömgren, B.: Astrophys. J. *108*, 242 (1948).
Swings, P., Rosenfeld, L.: Astrophys. J. *86*, 483 (1937).
Thaddeus, P., Clauser, J. F.: Phys. Rev. Lett. *16*, 819 (1966).
Thaddeus, P.: Ann. Rev. Astron. Astrophys. *10*, 305 (1972).
Townes, C. H., Schawlow, A. L.: Microwave Spectroscopy. New York: McGraw Hill 1955.
Townes, C. H., Cheung, A. C.: Astrophys. J. *157*, L 103 (1969).
Watson, W. D., Salpeter, E. E.: Astrophys. J. *174*, 321 (1972a).
Watson, W. D., Salpeter, E. E.: Astrophys. J. *175*, 659 (1972b).
Weinreb, S., Barrett, A. H., Meeks, M. L., Henry, J. C.: Nature *200*, 829 (1963).
Weinreb, S.: Nature *195*, 367 (1962).
Werner, M. W., Salpeter, E. E., in: Proc. of the 16th Liege Symposium, p. 113. Inst. d'Astrophysique de l'Universite de Liege 1970.
Whiteoak, J. B., Gardner, F. F.: Astrophys. Lett. *11*, 15 (1972).
Wilson, R. W., Penzias, A. A., Jefferts, K. B., Thaddeus, P., Kutner, M. L.: Astrophys. J. *176*, L 77 (1972).
Winnewisser, G., Maki, A. G., Johnson, D. R.: J. Mol. Spectrosc. *39*, 149 (1971).
Winnewisser, G., Winnewisser, M., Winnewisser, B. P.: MTP Rev. of Chemistry, Phys. Chemistry, Series 1, Vol. 3: Spectroscopy (ed. D. A. Ramsay). Oxford: Medical Technical Publ. Co., Ltd. and London: Butterworth Co., Ltd. 1972.
Winnewisser, G.: Private communication (1972a).
Winnewisser, G.: Paper 6B5, 2nd European Microwave Spectroscopy Conference, Bangor, Wales, U. K. 1972.
Witt, A. N., Johnson, M. W.: Astrophys. J. *181*, 363 (1973).
Witt, A. N., Lillie, C. F.: Astron. Astrophys. *25*, 397 (1973).
Woolf, N. J.: Int. Astr. Union, Symposium Nr. 52, Albany, New York 1972.
Zuckerman, B.: Symposium on Interstellar Molecules, Charlottesville, Va. 1972.
Zuckerman, B., Morris, M., Turner, B. E., Palmer, P.: Astrophys. J. *169*, L 105 (1971).

Received March 14, 1973.

# Carbon Chemistry of the Apollo Lunar Samples

**Prof. Geoffrey Eglinton, Dr. James R. Maxwell and Dr. C. T. Pillinger**

The Organic Geochemistry Unit, School of Chemistry, University of Bristol, Bristol, England

**Contents**

I.     Introduction . . . . . . . . . . . . . . . . . . 84

II.    Viable, Dead and Fossil Microorganisms . . . . . . . . 84

III.   Solvent Extractable Carbon Compounds . . . . . . . . . 85

IV.   Lunar Surface Processes . . . . . . . . . . . . . . 89

V.     Total Carbon Measurements . . . . . . . . . . . . . 90

VI.   Sources of Carbon on the Moon . . . . . . . . . . . 91

VII.  Mineralogical Studies . . . . . . . . . . . . . . . 92

VIII. Pyrolysis and Volatilisation Studies . . . . . . . . . . 92

IX.   Acid Dissolution/Hydrolysis Studies . . . . . . . . . 95

X.     Origins of the Hydrocarbons and Carbides in the Fines and Breccias . . . . . . . . . . . . . . . . . . . . . 97

XI.   Carbon Isotopes . . . . . . . . . . . . . . . . . 104

XII.  Carbon Chemistry as an Exposure and Reworking Parameter . . 105

XIII. Analytical Artifacts . . . . . . . . . . . . . . . . 106

XIV. Compounds of N, P and S . . . . . . . . . . . . . . 107

XV.  Conclusions . . . . . . . . . . . . . . . . . . . 107

XVI. References . . . . . . . . . . . . . . . . . . . . 108

## I. Introduction

The Apollo lunar samples have provided the first opportunity for the analysis of material collected from an extraterrestrial body under carefully controlled conditions. No definitive evidence about the concentration and nature of lunar carbon was provided by the $\alpha$ back-scattering experiments carried out by the Sureveyor unmanned landers. However, these remarkably successful experiments did indicate a magmatic origin for the samples analysed[1-3]. This was confirmed by the Apollo missions, three distinct samples types being collected from the lunar regolith:

1. A variety of igneous rocks of magmatic origin.
2. Breccias which are shock compressed aggregates of soil (regolith breccias) or thermally welded material (metamorphic breccias).
3. Fine grained soil comprising complex mixtures of mineral and glass fragments and microbreccias[4-7].

There was, therefore, little likelihood of complex organic compounds being present. However, a search for such compounds was included in the analytical schemes of most of the Bioscience Investigators[8]. Four basic approaches have been attempted to study the nature and distribution of carbon and its compounds in the Apollo samples:

1. A search for viable, dead, or fossil microorganisms.
2. A search for solvent-extractable compounds (hydrocarbons, etc.) of the type found in many terrestrial geological situations.
3. Pyrolysis and combustion experiments to determine the concentration and isotopic composition of total carbon, and pyrolysis experiments to detect polymeric material *via* its pyrolytic products.
4. Analysis of gaseous constituents released by *in vacuo* crushing or by dissolution of the inorganic matrix in concentrated mineral acid.

Each approach will be described in some detail and its relevance to the carbon chemistry of the moon discussed.

## II. Viable, Dead and Fossil Microorganisms

Prior to the release of any Apollo 11, 12 and 14 samples to the Principal Investigators, aliquots of the samples from each of these missions were screened at the Lunar Receiving Laboratory, Houston (LRL) for the presence of possibly harmful microorganisms. However, a wide variety of living organisms were exposed to the samples without any pathogenic effects. Addition of the samples to over 300 different nutrient media and environments also failed to reveal the presence of viable microorganisms, including terrestrial contaminants[4,5,9-11]. These results also showed that the moon had not been irreversibly contaminated by earlier probes. The only viable microorganisms obtained (*Streptococcus mitis*) were found in foam packing from within the Surveyor III TV camera which had resided on the lunar surface for two and a half years and was returned by the

Apollo 12 mission[12]. Some terrestrial microorganisms apparently have the potential of surviving on the moon for long periods of time in the protected interior environment of the camera.

Micropaleontological examination of rock chips, dust and thin sections by optical and microscopic techniques provided no indication of fossil or dead microorganisms[13-16]. The method detected, however, terrestrial contamination in the form of particulate organic matter such as shreds of teflon, silicone rubber and cellulose fibres[13].

## III. Solvent-Extractable Carbon Compounds

A search in the Apollo 11 fines for various classes of medium molecular weight biolipids and their derived fossil counterparts was carried out by almost all of the Bioscience Investigators. Isolation methods were based mainly on extraction with water or organic solvents with or without acid hydrolysis. Two isolation schemes are worth particular mention. Abell et al.[17] used a closed analytical system which allows extraction with a variety of solvents, evaporation and column chromatography to be carried out in an atmosphere of helium. Kvenvolden et al.[18] used a sequential scheme of solvent extraction and hydrolysis to analyse one sample of fines for a wide variety of compound classes (Table 1). The sample was retained in the same vessel throughout the procedure to minimize contamination through excessive handling of sample and extracts. Both of these schemes should have a wider applicability. For example, a combination of the two could provide the basis for remote automated analyses of planetary surfaces by unmanned landers[19]

The analytical methods used by most investigators were based on detection by gas chromatographic and mass spectrometric equipment operated at the highest sensitvities available. The findings are summarised in Table 1. No aliphatic or aromatic hydrocarbons, fatty acids, alcohols, esters, sugars, purines and pyrimidines (bound and free) were found at detection limits of a few parts per billion. The situation with respect to porphyrins and amino acids is less clear (see below). The net result, however, of the search for medium molecular weight compounds of biological interest was that there was no evidence for life, either present or past, on the moon. This result was expected by almost all of the investigators but the experiments had to be performed and the experience gained should prove invaluable in the analysis of other extraterrestrial samples, in which the presence of living organisms is a more likely possibility.

Rho et al.[27-29] have consistently claimed the absence of porphyrins in Apollo 11, 12 and 14 fines at detection limits corresponding to approximately 0.005 parts per billion. Hodgson et al.[30,31] detected porphyrin-like pigments in Apollo 11 fines and one sample of Apollo 12 fines collected near the Lunar Module. It was concluded that these contaminants arose from the Lunar Module descent rocket exhaust, similar combustion products having been observed in dunnite exposed in the laboratory to the exhaust. No porphyrins were found by either

85

Table 1. Search for carbon compounds in lunar fines by solvent extraction

| Compounds sought | Sample of fines | Isolation procedure | Analytical method | Present (+) or absent (−) | Approximate amount (Detection limit) if absent | Comments | Ref. |
|---|---|---|---|---|---|---|---|
| Volatile[1] compounds or derivatives | A11 | $C_6H_6$/$CH_3OH$ extraction and silylation of extract | GC–MS | − | 2 ppb | Closed extraction and derivatisation systems | 17) |
| Volatile[1] compounds | A11,A12 | $C_6H_6$/$CH_3OH$ extraction before and after HF dissolution | HRMS | − | 1 ppb | LM exhaust (A11 and 12) phthalate, silicone oil (A11) contaminants detected | 20, 21) |
| Volatile[1] compounds | A11 | (1) $C_6H_6$/$CH_3OH$ extraction; (2) $CH_2Cl_2$ extraction of HCl, HF dissolution products | GLC,GC–MS | − | 1 ppb | Variety of contaminants and artifacts detected. Possibly indigenous? | 22) |
| | | | HRMS | +(Aromatic hydrocarbons) | <1 ppb | | |
| Volatile[1] compounds amino acids (bound) | A11,A12 | (1) $C_6H_6$/$CH_3OH$ extraction; (2) HCl hydrolysis | (1) GLC; (2) GLC of derivatives | −; − | 1 ppb; 2 ppb | ? Isopropyl disulphide contaminant | 23, 24) |
| Volatile[1] and non polar compounds; free amino acids | A11,12 | (1) $C_6H_6$/$CH_3OH$ extraction (A11); (2) $H_2O$ extraction | TLC, GC–MS; Amino acid analyser | +(Hydrocarbons); + | few ppm (A11); up to 0.1 ppm | ? Probably contaminants (a phenol and a methyl ester also) | 25, 26) |
| Porphyrins | A11,A12, A14 | Various extraction procedures and demetallation | Spectrofluorometry | − | 0.005 ppb | (See text) | 27–29) |

| | Sample | Extraction method | Analysis | +(A11,A12) | Concentration | Remarks | Ref. |
|---|---|---|---|---|---|---|---|
| Porphyrins | A11,A12 A14 | C₆H₆/CH₃OH extraction and demetallation | Spectrofluorometry | +(A11,A12) | 0.1 ppb(A11) 0.05 ppb(A12) | LM exhaust contaminant (A11) (see text) | 30–32) |
| Alkanes (C₁₅ to C₃₀) | A11 A12 | C₆H₆/CH₃OH extraction before and after crushing and HF dissolution | GLC | – | 1 ppb | | 33–34) |
| Organic compounds, purines, pyrimidines | A11 | C₆H₆/CH₃OH extraction | NMR liquid/liquid chromatography (UV detector) | – – | 200 ppm 10 ppb | | 35) |
| Sugars amino acids (free and bound) (A11,A12,A14) hydrocarbons purines and pyrimidines fatty acids | A11,A12 | Extraction and Hydrolysis | Mainly GLC of derivatives | – | few ppb | For amino acids see text | 18, 36) |
| Amino acids (free and bound) | A11,A12, A14 | Aqueous extraction and hydrolysis | Amino acid analyser | + | up to 70 ppb | See text | 37–39) |
| Amino acids (free and bound) | A11,A12, A14 | Aqueous extraction and hydrolysis | GLC of derivatives | – | 1 ppb | Present in one A14 sample ppb concentrations | 40,41) |

1) Sufficiently volatile for gas-liquid chromatography and/or mass spectrometry.

investigator in a sample of Apolle 14 fines extracted by Hodgson *et al.*[32]. In a trench sample of Apollo 12 fines collected at a point well-removed from the LM, however, porphyrin-like pigments, which are thought to be indigenous to the sample, were detected. Hodgson *et al.* [31,32] believe that these compounds represent either laboratory oxidation products of lunar precursors or derive from extralunar sources such as micrometeorites, although Rho *et al.* [28] suggest that absorptions observed may be light scattering anomalies due to grating defects in the spectrofluorometer.

The controversy over the presence or absence of free or bound (released by acid hydrolysis) amino acids is indicated by the data in Table 1. Some investigators reported the absence of amino acids in Apollo 11 and 12 fines, whereas other reported part per billion concentrations, especially glycine (gly) and alanine (ala) (Fig. 1). Accordingly, a collaborative study between two la-

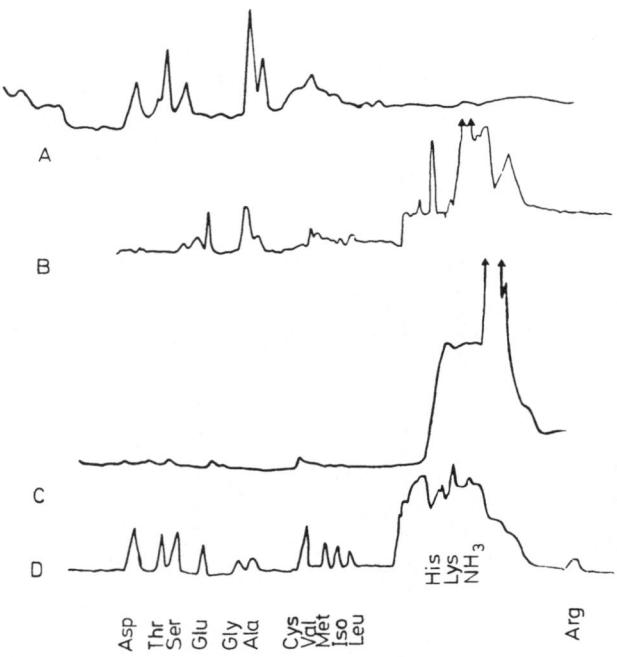

Fig. 1. Amino acid profiles (from amino acid analyser). (A) Hydrolysate of an aqueous extract from Apollo 11 bulk fines (sample 10086); basic amino acids – histidine, lysine, etc. – not shown. (B) Hydrolysate of an aqueous extract from Apollo 12 fines (Sample 12033). (C) Water blank carried through entire analytical sequence. (D) Standards

boratories was carried out to examine the Apollo 14 SESC fines (Special Environment Sample Container)[39,41]. This sample had been sealed on the moon (although not completely effectively) and was opened in a contamination-free facility at the University of California, Berkeley. Contamination of the sample was therefore minimal. The collaborative analysis indicated the presence of both free and bound amino acids at part per billion concentrations. The latter are thought to represent not bound amino acids but the products of acid hydrolysis of organic precursors indigenous to the samples[39]. It has been suggested[42] that these simple precursors, which react together on acid hydrolysis, may be ammonia (from hydrolysis of nitrides), simple cyanides such as HCN, and aldehydes such as formaldehyde. The aldehydes (or carbon at the same oxidation state) could derive from hydrolysis of carbide, or hydration of acetylenes derived from hydrolysis of carbide, or from the reduction and hydrolysis of cyanides[42]. There is evidence that the proposed parent species (nitrides, cyanides and carbides) are present in lunar fines. Dissolution of a number of samples of Apollo 14 fines in DF released DCN, indicating that cyanide species are present[43]. Indirect evidence for the presence of nitrides is given by the pyrolytic release below 600 °C of $NH_3$ from the fines, probably from the reaction of $H_2O$ with nitrides[43] and by the generation of ammonium species on acid hydrolysis[44]. *Carbides* are almost certainly present in a wide variety of Apollo 11, 12, 14 and 15 samples[43,45-50].

If the amino acids released by acid are not contaminants, proof of their origin would be a fascinating, if difficult, experimental problem. Laboratory simulation studies, involving implantation of carbon at solar wind energies into metal targets, suggest that carbide can be synthesized on the lunar surface by solar wind implantation[51]. In addition, it seems likely that nitride and cyanide species should be synthesized by implantation[43,44]. A valuable experimental study would be hydrolysis of samples, previously irradiated consecutively in the laboratory with the appropriate ions (C, N, O, H) at solar wind energies, followed by a search for amino acids in the hydrolysate[52].

## IV. Lunar Surface Processes

To date lunar carbon chemistry studies have been primarily concerned with analyses of the fine material and, to a lesser extent, breccias. A brief description of the processes leading to the formation of these samples in the regolith is therefore necessary in order to discuss their carbon chemistry. Successive impacting meteorites generated sudden blanket depositions of layers of ejecta. Accompanying base surges of hot clouds of dust and gas, and shock waves, are thought to have resulted in compaction to form breccias. On a smaller scale micrometeorite impact, sputtering by cosmic particles and abrasion lead to further comminution and the formation of microbreccias. Micrometeorite impacts also cause the slow turnover (gardening) of the regolith to a depth of 2–3 cm, although other mechanisms, including electrostatic effects and downslope migration, must also contribute to this reworking. As a result, fine material is exposed to the flux of the solar wind and cosmic rays, and to any vapour phase generated by impact.

## V. Total Carbon Measurements

The total carbon contents of a variety of fines, breccias and crystalline rocks have been measured by a number of investigators[54−63]. Samples were either combusted directly in an oxygen atmosphere to $CO_2$ or first heated *in vacuo* and the gases released converted to $CO_2$. Fig. 2 summarises the data obtained by Moore *et al.*[7, 54−56] for Apollo 11 to Apollo 15 samples and the following general conclusions may be drawn[64]:

1. The highest values measured (*ca.* 200 ppm and higher for fines and regolith breccias and 60 ppm for crystalline rocks) are probably indicative of significant terrestrial contamination.

2. The fines and breccias have concentrations of carbon which are, in general, higher than those in the igneous rocks, although there are some exceptions, for example the fines collected near Cone Crater by the Apollo 14 mission, and the metamorphic breccias (see below).

3. The rocks are substantially depleted in carbon, in view of the relative cosmic abundance of this element but it is not known whether this is a result of volatilisation losses during melting or whether the moon has always been depleted in this and other volatile elements.

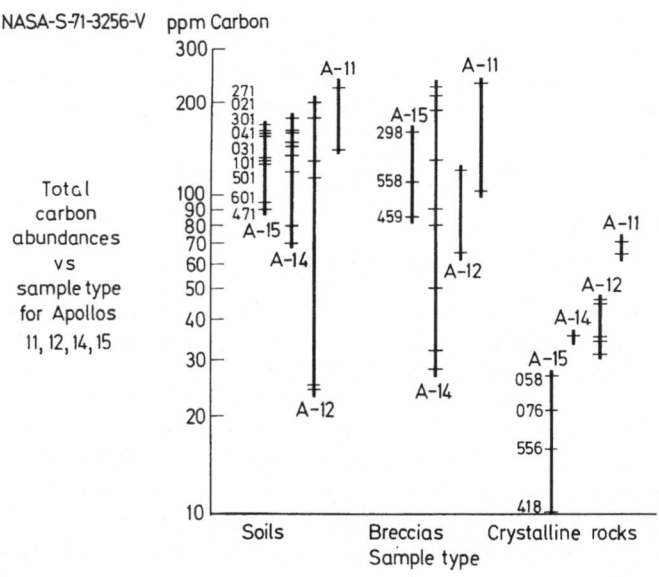

Fig. 2. Total carbon abundances (by combustion) in basaltic rocks, breccias and bulk fines from Apollo 11, 12, 14 and 15 missions. Horizontal bars indicate different samples

These data show that, because the fines derive from crystalline rocks, a carbon component has been added to the fines. The total carbon concentration in Apollo 11 fines increases with decreasing particle size, indicating that there is a surface correlation[54, 62]. It has been suggested[54] therefore, that solar wind implantation of carbon into the surfaces of the fines makes a significant contribution to the total carbon, although a meteorite contribution cannot be discounted.

## VI. Sources of Carbon on the Moon

There are probably three major sources which contribute carbon to the lunar surface:

*First*, primordial or magmatic carbon trapped during the crystallisation of the igneous rocks: the carbon content measured for the rocks should be representative of carbon from this source.

*Second*, the flux of meteorites and micrometeorites impacting on the lunar surface might be expected to contribute carbon in some form. In particular, iron meteorites contain small concentrations of the carbide mineral, cohenite [(Fe,Ni)$_3$C], graphite and diamond[65]. Carbonaceous chondrites contain up to 3.5% carbon as carbonate, polymeric material and solvent-extractable organic compounds[53]. Neutron activation analysis of the fines, breccias and crystalline rocks indicates that the regolith breccias and fines are enriched in a number of volatile and siderophile trace elements including Ir, Au, Cd and Bi. The enrichment is thought to represent a micrometeorite contribution of carbonaceous chondrites of up to 2%[66-68]. This contribution alone would provide up to 700 ppm of carbon, a figure significantly in excess of the observed total carbon contents of the fines and breccias. Losses of carbon through volatilisation on impact are evidently extensive. Laboratory simulation studies indicate that the temperature of the vapour cloud generated by the impact of a micrometeorite could be as high as 2000 °K[69].

*Third*, a contribution is likely from implantation by solar wind ions. When the moon is outside the earth's magnetosheath these species are implanted into unshielded surfaces. Carbon is the fourth most abundant element in the solar wind (*ca.* 4 x 10$^4$ nuclei/cm$^2$/sec$^1$) after hydrogen (*ca.* 3 x 10$^8$ nuclei/cm$^2$/sec$^1$, helium, and oxygen[70]. The carbon which could have been contributed to the fines from this source has been calculated from the content of certain rare gases in the fines and the relative abundance of carbon and the rare gases in the solar wind. Estimates range from 1 ppm (based on He) up to 50 ppm (based on Kr) and 300 ppm (based on Xe)[54, 71]. The descrepancies probably result from preferential diffusion of the lighter rare gases. Unfortunately the rates of diffusion for carbon or the rare gases are not known and the calculations are of restricted value at present. The depth of penetration into exposed surfaces by ions at solar wind energies (*ca.* 1 Kev/nucleon) would be *ca.* 1000 Å and one consequence is that carbon in the fines from this source would be expected to be correlated with the surface areas of the grains. This relationship alone would be insufficient, however, to define a solar wind origin because any carbon deposited by condensation as a result of meteoritic impacts would also be surface correlated. Other factors must be taken into consideration to distinguish these two sources.

In comparison with the solar wind flux (*ca.* $10^8$ particles/cm$^2$/sec), contributions to lunar carbon from implantation of solar flare and galactic cosmic rays will be minimal. The total flux of solar cosmic rays is *ca.* $10^2$ particles/cm$^2$/sec and that of galactic cosmic rays *ca.* 1 particle/cm$^2$/sec. These high energy particles, however, might produce spallogenic carbon species.

One other mechanism has been suggested[71] which might involve recycling of carbon already in the samples or in meteorites. Carbon species vapourised and contributed to the tenuous atmosphere as a result of impact, are expected to become ionized by interaction with solar wind ions or ultraviolet radiation, (then accelerated by the electric fields present), and subsequently implanted like solar wind ions, but at lower energy. At present, however, there is no proof that this mechanism does operate.

Definition of the ultimate origin of the carbon in the samples will certainly be difficult, if not impossible, because the processes occurring at the lunar surface must inevitably lead to some loss of identity. This is particularly applicable to any mechanism involving meteorite impact.

## VII. Mineralogical Studies

There are few reports of the presence of carbon-containing minerals in lunar samples. The carbide, cohenite has been observed as a trace component in Apollo 11 fines[72-74] and in one Apollo 14 breccia[75]. These grains probably represent meteoritic material not vapourised on impact; the carbide observed in the Apollo 14 breccia has the same elemental composition as the cohenite in the Odessa meteorite[75]. One grain of graphite (*ca.* 2 mm) was found in Apollo 11 fines[76] and there is one report of the presence of a grain of aragonite ($CaCO_3$) in a sample container used to transport an Apollo 11 breccia, although this is probably a terrestrial contaminant[77]. At present mineralogical examination of lunar material has provided little information about the carbon chemistry of the moon.

## VIII. Pyrolysis and Volatilisation Studies

Pyrolysis studies played an invaluable role in estimating the concentrations of organic contamination in the samples. Before the Apollo 11 mission it was realised that the concentrations of indigenous organic matter in the samples would be very low, or zero. Contamination of the samples with organic compounds of terrestrial origin, at all stages of the collection and subsequent handling procedures, was therefore thought to represent a major problem. A pyrolysis-low resolution mass spectrometer computer system was used at the LRL to monitor the procedures used for collecting and processing the samples[4, 5]. As a result, the levels of contamination in blank samples processed at Houston were decreased from *ca.* 1000 ppm before Apollo 11 to $<0.1$ ppm before Apollo 12. The contaminants detected have been reviewed[78] and will not be discussed in detail here. Major sources of contamination included the sample processing vacuum chamber and glove boxes at Houston, and the rocket exhaust of the lunar module.

Lists of possible contaminants were made available to the Bioscience Investigators who were subsequently able to recognise some of the species observed in the samples. Thus, pyrolysis-mass spectrometric analysis of Apollo 11 and 12 fines indicated the presence of contaminants including ethylene glycol, phthalate esters, and hydrocarbons up to mass 250[22, 79]. The LRL system was also used to provide a rapid measure of the total concentration of organic matter in the Apollo 11 and 12 samples[5, 6]. The quantities measured did not exceed 1 part per million, and even a significant proportion of this may have been contamination or artifactual aromatic hydrocarbons (see below). The most striking observation made by almost all of the Bioscience Investigators, however, is that the most abundant gaseous species released (Fig. 3) on vacuum pyrolysis, up to and beyond the melting point (*ca.* 1100 °C), are CO and $CO_2$[80] (Fig. 3). Taken together, these species account for $>90\%$ of the total carbon present in the fines. However, acid dissolution experiments (see below) indicate that they are not present as such, except perhaps as trace constituents or as adsorbed terrestrial contamination in the case of $CO_2$.

The CO is released mainly above 700 °C[23, 43, 80–82] and is thought to be a reaction product of carbon-containing phases, such as carbide, with the metal oxides and silicates present[24, 48, 81, 83]. In fact, Gibson and Johnson[81] have obtained a similar CO temperature release profile from synthetic mixtures of fines comprising olivine, pyroxene and anorthite, with traces of troilite, cohenite and graphite added. At temperatures above the melting point of the fines, small sporadic "bursts" of gas identified as CO or $N_2$ by low resolution mass spectrometry (Fig. 3) are thought to arise from outgassing of the melt or the rupture of inclusions[80].

$CO_2$ is released over two temperature ranges (below *ca.* 600 °C and above *ca.* 800 °C). The low temperature $CO_2$ (and $H_2O$) has been ascribed to adsorbed terrestrial contamination because samples of fines exposed to the laboratory atmosphere, before and after pyrolysis, showed similar temperature release patterns over the low temperature range[24, 80]. In this respect, one inference made from solar wind implantation simulation studies is noteworthy[51]. Isotopically-labelled carbon ions were implanted into lunar fines at energies similar to those calculated for the solar wind. Pyrolysis to 800 °C of the implanted targets failed to reveal the presence of any $CO_2$ containing isotopically-labelled carbon. It was inferred that the carbon in $CO_2$ released from the fines over the low temperature range does not derive from the solar wind. Hayes[71] has suggested that some of the carbon in this $CO_2$ could derive from vapourised species contributed to the lunar atmosphere from meteorite impact and subsequent implantation into the lunar surface. At present there is no evidence for this hypothesis.

93

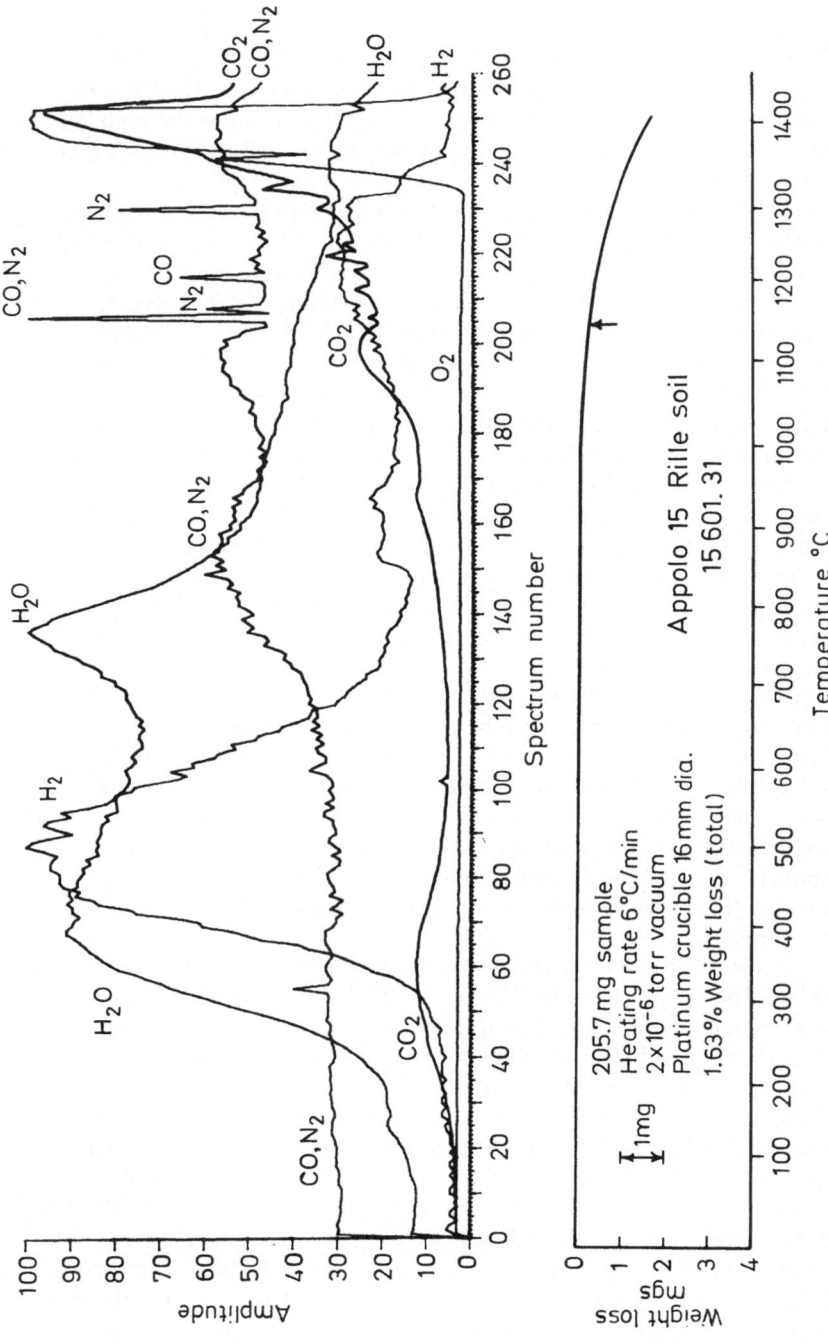

Fig. 3. Temperature-gas release profiles (computerised mass spectral output) and weight loss profile from pyrolysis of an Apollo 15 sample of fines

The high temperature $CO_2$ is certainly pyrolytic in origin and may derive from reactions similar to those which are thought to synthesise the $CO$[81]. In the intermediate temperature range low molecular weight compounds, including $CH_4$, HCN, $CS_2$ and simple aromatic compounds, including benzene and toluene, are evolved at concentrations up to 2 $\mu g/g$ of individual species[43, 84]. It has been suggested that they are synthesised by the thermally-induced reaction of solar wind products in the surface layers of the particles of fines[43]. Pyrolysis in a helium atmosphere (760 Torr), followed by GC-MS analysis of the products, affords somewhat more complex species than observed by vacuum pyrolysis[25, 26, 84]. A variety of alkylbenzenes and alkylnapthalenes thiophene, indene, and biphenyl, at a total concentration of about 25 ppm, are released. It is possible that these represent secondary synthetic products, possibly from the reaction of the CO and $H_2$ released, catalysed by the lunar fines[84]. These secondary reactions occur to a greater extent in the high pressure pyrolysis system; this was tested by addition at 700 °C of CO and $H_2$ to a previously pyrolysed lunar sample, traces of benzene being synthesised.

## IX. Acid Dissolution/Hydrolysis Studies

Perhaps the most detailed information about the carbon chemistry of lunar samples has been obtained from experiments, conducted under vacuum, which

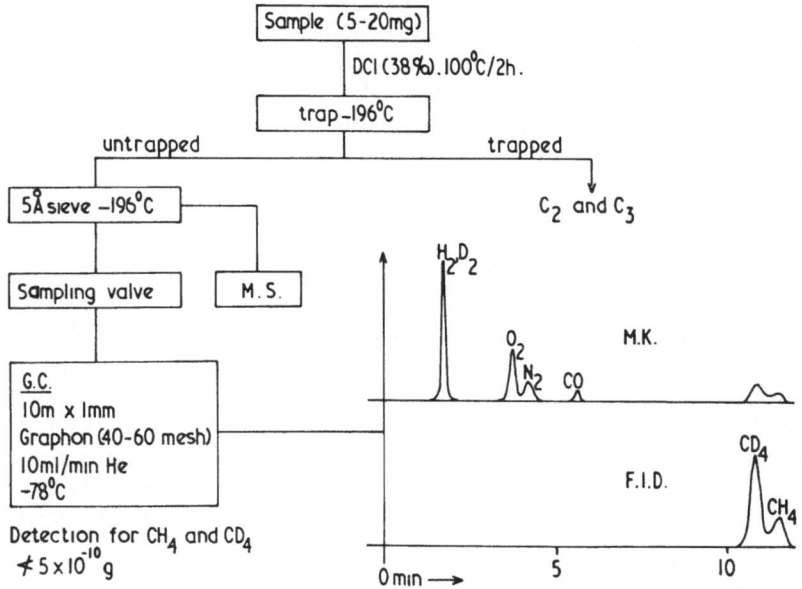

Fig. 4. Schematic of one method for analysis of carbide and $CH_4$ in lunar samples by deuterated acid dissolution and gas-solid chromatography

involve dissolution of the mineral matrix in concentrated mineral acids, followed by analysis of the gaseous species released.

A typical analytical scheme[47] is shown in Fig. 4. After removal of adsorbed gases from the sample (5–20 mg) and of dissolved atmospheric gases from the acid, the sample and acid are allowed to come into contact and react. Gases released are concentrated by trapping into two fractions, *viz*, those condensed at − 196 °C and those condensed on 5 Å molecular sieve at −196 °C. The two fractions are separately analysed by gas-solid chromatography on graphitized carbon (or by low resolution mass spectrometry, MS, in some cases). Detection of the separated components is achieved by coupling a microkatharometer detector (MK) and a flame ionization detector (FID) in series. The microkatharometer responds to species not measured by the flame ionization detector. Similar analytical systems have been used by Henderson *et al.*[21] and Chang *et al.*[48, 83]. High resolution mass spectrometry[20, 43, 48, 50] and combined gas chromatography mass spectrometry, with multiple ion plotting,[23, 24, 85] have also been used to analyse the released gases.

Dissolution of lunar fines with HF or HCl releases as the major products the $C_1$ to $C_4$ alkanes, $C_2H_2$ and $C_2H_4$ in concentrations which represent up to *ca.* 20% of the total carbon present. The major component by far is methane, which may be released in concentrations up to *ca.* 25 $\mu$g/g as carbon. The use of deuterium labelled acids for dissolution allows distinction between hydrocarbons present as such and those produced by acid hydrolysis[43, 45–50, 85, 86].

Thus methane is released from the fines mainly as $CD_4$ and $CH_4$ in ratios ($CD_4 : CH_4$) varying from 2.2 to 12.7: the most common ratio being in the range 3 : 1 to 5 : 1. Some is also released as $CD_3H$ but is considered to be an acid hydrolysis product and is included with $CD_4$ for quantitative measurement. Other partly deuterated species ($CH_2D_2, CH_3D$) are released in only trace concentrations. Similarly, ethane is released as both undeuterated and deuterated (and almost fully deuterated) species. Other species (ethylene, acetylene, propane, propene, butane and butene) appear to be mainly deuterocarbons. The $C_1$ to $C_4$ deuterocarbons are thought to arise from acid hydrolysis of carbides in the samples (Table 2).

Only trace concentrations of CO are released ($< 3$ $\mu$g/g) and even this may be an artifact. The $CO_2$ observed may be an adsorbed terrestrial contaminant, although it has been suggested[85] that it could be a hydrolysis product of carbonate in the samples. The high concentrations of CO and $CO_2$ observed on heating the fines are pyrolytic products. Other carbon-containing species released by dissolution include DCN, indicating the presence of cyanide species, and traces of $CS_2$, possibly from implantation of solar wind carbon into troilite, FeS (c.f. origin of hydrocarbons, below).[43] The latter would also be expected to give rise to $D_2S$ on hydrolysis. $D_2S$ has been shown to exchange during gas

$$M_xC_y(\text{carbide}) \xrightarrow{\text{DCl or DF in } D_2O} CD_4, C_2D_4, C_2D_6 \dots \text{ etc.}$$

$$CH_4, C_2H_6, \text{ etc. (trapped gas)} \xrightarrow{\text{DCl or DF in } D_2O} CH_4, C_2H_6 \dots \text{ etc.}$$

chromatography (Table 2), only $H_2S$ being detected. Indigenous $H_2S$ gas and troilite hydrolysis products are therefore indistinguishable[86].

Table 2. Carbon-containing and other gaseous species released by *in vacuo* deuterated acid (DF or DCl in $D_2O$) dissolution of lunar fines

| Species | Identified by | Comment | Ref. |
|---|---|---|---|
| $CH_4,C_2H_4,C_2H_6$ | $GC^1)$,$LRMS^2)HRMS^3)$,$GCMS^4)$ | Solar wind derived | 24,43,46–50 85,86) |
| $CD_4$ | GC, LRMS, HRMS, GCMS | Carbide hydrolysis product, correlates with lunar surface exposure | 24,43,46–50 85,86) |
| $C_2D_2,C_2D_4,C_2D_6,$ $C_3D_6,C_3D_8,C_4D_8,$ $C_4D_{10}$ | HRMS, GC | Carbide hydrolysis products | 43,48–50,86) |
| CO | GC, GCMS | Mainly an artifact; indigenous component small, if present. | 47,48,86) |
| $CO_2$ | GC, GCMS | Mainly adsorbed contaminant; indigenous component small, if present? Possibly reaction products from carbonate? | 48,85,86) |
| DCN | HRMS | Cyanide groups in fines | 43,50) |
| $CS_2$ | HRMS | Solar wind implanted carbon in troilite (FeS)? | 43,50) |
| $H_2S$ | GC, GCMS | Indigenous gas or troilite hydrolysis product; impossible to distinguish because of exchange (D and H). | 85,86) |

[1]) Gas solid chromatography.
[2]) Low resolution mass spectrometry.
[3]) High resolution mass spectrometry.
[4]) Combined gas chromatograph-mass spectrometry.

## X. Origins of the Hydrocarbons and Carbides in the Fines and Breccias

Before considering the available experimental evidence and the origins of these species in detail, it is necessary to examine possible origins in general terms. First, the origin may be lunar or extralunar; a lunar origin indicating species trapped during crystallisation of the magma. One likely extralunar origin would be solar wind implantation into the outer 1000 Å of exposed surfaces. Thus, surface-located $CH_4$ could be formed by a number of mechanisms, including hydrogenation of carbon in the fines by solar wind protons (see below). Similarly carbide, or material behaving as carbide on acid hydrolysis, could be formed by implantation of carbon into metal in the fines. The other likely extralunar origin would be from meteorites. Carbonaceous chondrites, however, contain lower concentrations of $CH_4$ than do the lunar fines and, in any case, losses on impact of this gaseous component would be high. A direct contribution to the $CH_4$, in the fines from the $CH_4$ in meteorites can therefore be ruled out [46, 47, 86].

Although only a few crystalline rocks have been analysed by the acid dissolution method [46, 86] or by crushing [87] the concentrations of $CH_4$ and $C_2H_6$ present are low. Also $CD_4$ is either absent or present in trace quantities in the dissolution products indicating that there is little or no carbide in the rocks. Hydrocarbons and carbide have therefore been added to the fines, indicating an extralunar origin for both species [46, 86]. The small quantity of $CH_4$ in the crystalline rocks is probably magmatic in origin. There may therefore be a small primordial component in the fines, although the extent of losses during the formation of the fines is unknown.

If the hydrocarbons and carbide in the fines have an extralunar origin, there should be a correlation of both species with exposure on the lunar surface. Abell et al. [86] and Cadogan et al. [46, 47] have attempted to correlate the concentrations of $CH_4$ and $CD_4$, representative of the gaseous hydrocarbons and carbide, respectively with physical and chemical parameters indicative of

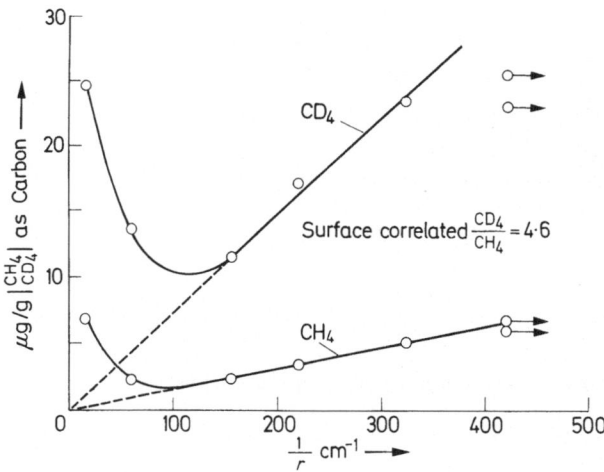

Fig. 5. Concentrations of $CD_4$ and $CH_4$ (as μg/g carbon) released by DCl dissolution plotted against the reciprocal of the mean grain radius of sieved fractions (Apollo 11 bulk fines)

this surface exposure. A general trend is evident between the $CH_4$ and $CD_4$ and the modal analysis data related to the degree of regolith reworking. The median grain size, sorting index, and the proportions of glass, microbreccias and glassy aggregates in the fines provide a general indication of the maturity of different samples [88, 89]. Samples showing the greatest degree of reworking have the highest $CH_4$ and carbide contents.

The $CH_4$ and $CD_4$ concentrations are directly proportional (Fig. 5) to the reciprocal of the mean grain radius for sieved fractions of Apollo 11 fines for the range 48 $\mu$ to 152 $\mu$ mean grain diameter [47]. The coarser particles ($> 152 \mu$) show methane concentrations in excess of the surface related component, indicating that there is a volume related component which increases with grain size. This effect has also been observed in Apollo 12 fines for solar wind rare gases [90]. The volume related component probably arises from the presence of particles with composite grain surfaces (microbreccias and glassy aggregates). There is no evidence available at present, however, to explain the increase of the volume related component with particle size. Holland et al. [49] have also observed a surface area relationship from an examination of the gases released by DF dissolution of a sample of Apollo 14 fines. $CH_4$ was linearly related to the reciprocal of the radius for particles from 37 $\mu$ to 420 $\mu$ diameter. The $C_2H_6$ released was dependent on a higher power of the radius suggesting a bimolecular synthetic process. Further evidence for a surface location for $CH_4$ is indicated by shallow etching of the fines with NaOD [46,86]. Similarly, etching with DF vapour releases $CD_4$, which is consistent with surface-located carbide [43].

The lunar fines and regolith breccias contain a variety of isotopes of rare gases including $^{36}Ar$, whose presence is generally attributed to solar wind implantation. Fig. 6 indicates that there is a correlation between both the concentrations of $CH_4$ and $CD_4$ released by acid dissolution [47] of a variety of fines and regolith breccias and the measured content of $^{36}Ar$ [5-7, 90-95] a high concentration of $CH_4$ and $CD_4$ being associated with a high $^{36}Ar$ concentration. Holland et al., [43,50] have observed similar correlations between $CH_4$ (and $CH_4 + C_2H_6$) and $^{20}Ne$ and $^{36}Ar$.

Particles with very high solar flare cosmic ray track densities ($\geqslant 10^8$ tracks $cm^{-2}$) have a high probability of having experienced unshielded surface exposure during their history, that is, a high probability of exposure to the solar wind [76]. Fig. 7 shows a plot of $CH_4$ concentrations [47] for a number of samples of fines against the fraction of grains having these high track densities in each particular sample [76, 96-99]. Samples showing a high proportion of grains with high track densities also have high $CH_4$ concentrations. The proportions of grains with these track densities measured by Bhandari for the Apollo 14 samples are consistently lower (Fig. 7) than those measured by other investigators for the same samples. The relationship shows a tendency towards track density saturation as would be expected. The $CH_4$ concentrations continue to increase, whereas the proportion of grains with track densities greater than $10^8$ tracks $cm^{-2}$ cannot exceed 100%. Optical counting methods do not allow resolution of track densities greater than this figure; clearly a more accurate estimate of

99

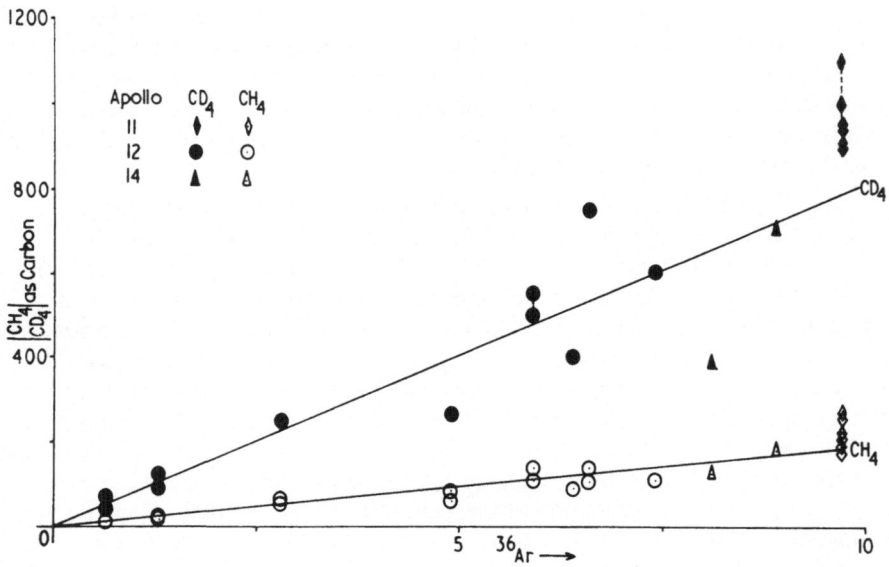

Fig. 6. Concentrations ($10^{15}$ atoms/g) of $CD_4$ and $CH_4$ (as carbon) released by DCl dissolution versus concentration of solar wind implanted $^{36}Ar$ ($10^{15}$ atoms/g) in Apollo 11, 12 and 14 fines

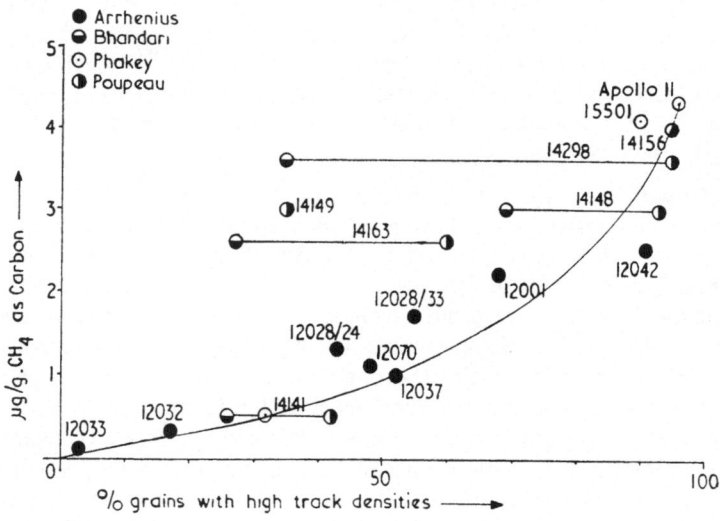

Fig. 7. $CH_4$ as carbon (μg/g) in total sample versus percentage of mineral grains with track densities $> 10^8$ cm$^{-2}$

unshielded surface exposure is the proportion of particles with extremely high track densities[100] ($> 10^{11}$ tracks $cm^{-2}$). These high densities can be measured by high voltage electron microscopy but at present insufficient measurements are available for useful correlation with $CH_4$ concentration. Again the $CD_4$ concentrations show a similar trend but there is a greater degree of scatter.

Electron microscopic examination of the finest particles (smaller than 400 mesh) show the presence of mineral grains having a coating, *ca.* 1000 Å in thickness, which is amorphous in comparison with the crystalline interiors of the grains[100-102]. This radiation damage is thought to be the result of solar wind bombardment and, in fact, the amorphous coatings can be generated in the laboratory by bombarding samples with rare gas ions at solar wind energies[102]. Samples showing the highest numbers of fine particles with amorphous coatings[103] have the highest $CH_4$ concentrations[47] (Fig. 8). This provides further indirect evidence for a solar wind origin for the hydrocarbons in the fines. Again there is a similar but less marked trend for the $CD_4$ concentrations[47].

The experimental evidence points, therefore, to a solar wind origin for the $CH_4$, and presumably the other gaseous hydrocarbons, in the fines. Information is also available from laboratory studies involving implantation of $^{13}C^+$ and $D_2^+$ at solar wind energies into samples of lunar fines and synthetic analogues containing the major lunar minerals[51]. The implanted targets were dissolved in HCl and the gaseous products analysed. Methane was released as $^{13}CD_4$, indicating a synthesis from implanting species alone. Methane as $^{12}CD_4$ was also released, resulting from deuterium and carbon ($^{12}C$) already present in the

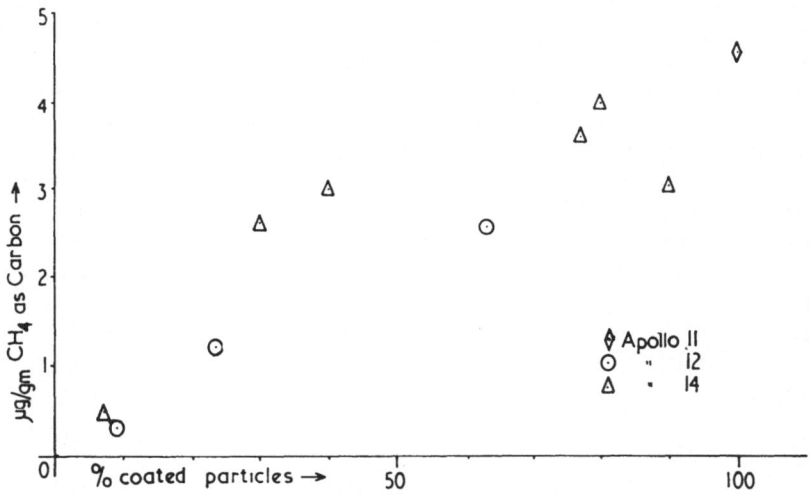

Fig. 8. Concentrations of $CH_4$ as carbon ($\mu$g/g) in bulk fraction versus percentage of grains of $1-2\mu$ diameter in the 400 mesh residue having amorphous coatings for Apollo 11, 12 and 14 fines

targets in some form. By analogy methane could be synthesised at the lunar surface by hydrogenation of solar wind implanted carbon or carbon already present from another source. Carbon of meteoritic origin might therefore contribute indirectly to the methane in the fines. A third mechanism would be the generation of methane by *in situ* hydrolysis of carbide on the lunar surface by water or hydroxylic groups synthesised within the matrix by the solar wind[17, 45]. At present there appears to be little chance of distinguishing experimentally between the relative importance of these mechanisms for the synthesis of solar wind generated $CH_4$.

The situation with respect to the origin of the carbide is more complex. The occasional grains of cohenite in a few of the samples of fines are almost certainly meteoritic in origin, so some carbide analysed as $CD_4$ must have been contributed in this 'form'. Although discrete grains are extremely rare, acid dissolution shows that carbide (or material behaving as carbide) is ubiquitous in the fines. The carbide, therefore, appears to be finely disseminated.

The surface area relationship of carbide (as $CD_4$) in the fines and the relationships with lunar surface exposure parameters are consistent with both a solar wind and a meteoritic origin. Again, simulation studies show that both mechanisms are possible[51]. First, implantation of $^{13}C^+$ at solar wind energies into metal films (Fe and Al) followed by acid dissolution in HCl results in the release of $^{13}CH_4$. A similar process could occur on the lunar surface by implantation of carbon ions into the abundant iron mounds and blebs observed[104–106] on the surfaces of the particles in the fines. "Carbide" generated in this way at submicroscopic levels would not be distinguishable by mineralogical techniques, but would be by chemical analysis employing acid hydrolysis as for discrete carbide grains. Second, vacuum vapourisation in the laboratory of iron filaments containing carbide gives rise to iron and carbon species which, when deposited, react as carbide on acid hydrolysis. Impacting iron meteorites could thus contribute "carbide" to the lunar surface. The quantities of carbide contributed directly from iron meteorites as discrete mineral grains, or through vapourisation, and subsequent deposition of carbide, are likely to account for only a minor porportion of the observed carbide in the fines. Calculations[47] based on the *maximum* carbon content of iron meteorites and on the unallowable assumption that *all* of the metallic iron in the fines derives from iron meteorites suggest that the *maximum* contribution to lunar carbide would be 10 $\mu$g/g of carbon as $CD_4$. This figure, however, assumes a 100% yield of $CD_4$ from acid hydrolysis. Hydrolysis in the laboratory of cohenite of meteoritic origin results in only a 6% yield of carbon as $CD_4$. A similar yield from the calculated direct contribution to lunar fines would account for only 0.6 $\mu$g/g of carbon as $CD_4$. In addition, the simulation experiments described indicate that solar wind implantation does not account for the remainder of the carbide in the fines. A third mechanism has been suggested for carbide synthesis[47]. Meteorites in the form of carbonaceous chondrites could make a major *indirect* contribution to lunar carbide. The iron in the fines is associated with the glass and is in an extremely finely divided form[106, 107]. Such an iron phase could arise from meteorite impact heating under reducing conditions. Dissolution of carbon already in the fines or from the meteorite *via* the vapour cloud formed in the iron, could give rise to carbide. Unfortunately, the surface area relationship for carbide in the fines and the correlation with lunar surface exposure, do not distinguish between a solar wind origin or the different mechanisms for a meteorite origin. Micrometeorite impact is a major process in reworking the regolith and hence, in exposing fresh fines to the solar wind. Contributions to carbide from meteorites and the solar wind should therefore increase in parallel. Correlation with a parameter indicative of the micrometeorite contribution does not distinguish between these origins. The carbide concentration[51] tends to increase with the concentration of a number of volatile and siderophile elements (Ir, Au, Se, Ag, Bi)[67, 108] indicative of the micrometeorite contribution.

A high concentration of *e.g.* Bi (Fig. 9) indicates an extensive micrometeorite contribution, but not necessarily of carbide. The relative contributions to lunar carbide from the solar wind and meteorites are therefore unknown. Studies involving magnetic separations of metal phases should prove informative. A crude magnetic concentrate of a sample of Apollo 14 fines has been shown to be enriched in carbide[43]; a number of fractions obtained from an Apollo 11 sample by density and magnetic susceptibility separations show widely varying concentrations of carbide[109]. Examination of the trace elements in the metal associated with the carbide should allow distinction between meteorite and non-meteorite metal.

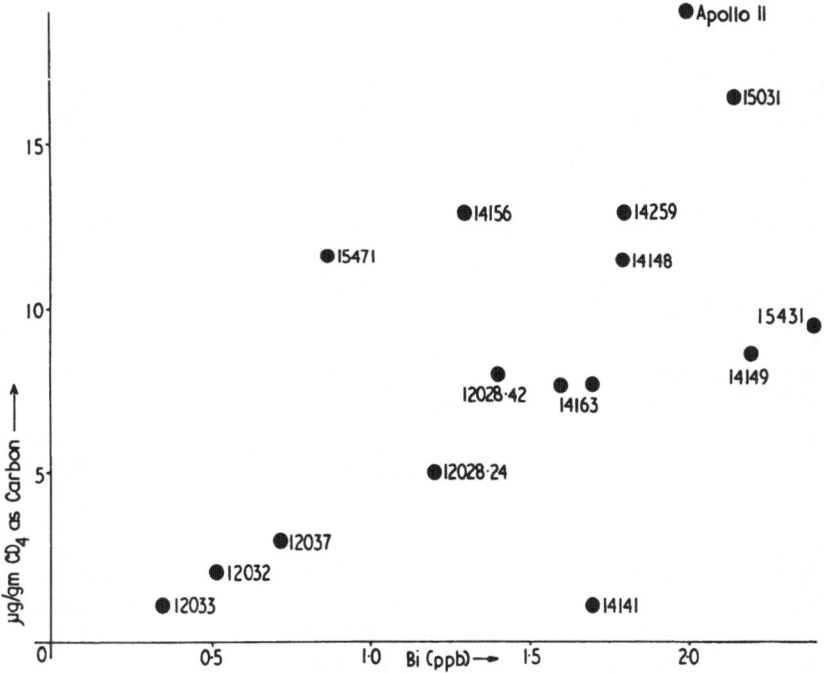

Fig. 9. Concentration of $CD_4$ as carbon released by DCl etch of samples of Apollo 11, 12, 14 and 15 fines versus bismuth concentrations

## XI. Carbon Isotopes

Stable carbon isotope ratios have been measured for a variety of samples of each type. Data are quoted as $\delta^{13}C$ values relative to the PDB standard (Pee Dee Belemnite) where

$$\delta^{13}C \, ^{0}/oo = \frac{^{13}C/^{12}C \text{ sample} - ^{13}C/^{12}C \text{ standard}}{^{13}C/^{12}C \text{ standard}} \times 1000.$$

Values for the few basaltic rocks examined fall in the range $-20^{0}/oo$ to $-25^{0}/oo$, similar to those reported for terrestrial basalts[61-63]. The one metamorphic breccia examined has a value of $ca. -18^{0}/oo$, tending towards those of the igneous rocks[110].

The fines and regolith breccias consistently show[57-59, 62, 63, 110] an enrichment in $^{13}C$, ranging from $-3.6^{0}/oo$ to $+20.2^{0}/oo$. It should be noted that where more than one analysis of the same sample is available the higher value is considered[62] more accurate because it is presumed less likely to have been affected by isotopically light terrestrial contamination. In fact, Kaplan[64] has shown that, for a series of samples of Apollo 11 bulk fines, the sample having the lowest total carbon content (least contaminated) has the highest $\delta^{13}C$ value. One investigator has consistently reported values for the fines and breccias which are low and possibly reflect terrestrial contamination[60, 61].

The enrichment in the fines and regolith breccias is thought to result from exposure on the lunar surface, resulting in either preferential removal of $^{12}C$ or addition of $^{13}C$. A number of mechanisms have been proposed to explain these observations for the fines and must also be applicable to the regolith breccias.

1. Preferential removal of $^{12}C$ occurs by a "hydrogen stripping" process; volatile carbon species such as $CH_4$ being produced by interaction with protons and subsequently lost[62, 111]. The fundamental requirement for this explanation, i.e. synthesis of volatile species by proton irradiation at solar wind energies, has been demonstrated by Pillinger et al.[51]. Also there appears to be a correlation between $CH_4$ concentration and the $\delta^{13}C$ value of total carbon; a high $CH_4$ concentration being associated with a high $\delta^{13}C$ value[46].

2. Selective volatalisation[57] or condensation[47] occurs as a result of meteorite impact ($^{12}C$ species are more volatile than $^{13}C$ species). A similar mechanism has also been proposed[57] to explain the enrichment of $^{30}Si$ over $^{28}Si$, and $^{18}O$ over $^{16}O$.

3. Mixing involving fines of igneous origin ($\delta^{13}C-25^{0}/oo$), and solar wind carbon and meteorite debris of assumed high $\delta^{13}C$ values[59]. This mechanism is consistent with the relationship (Fig. 10) between total carbon and $\delta^{13}C$ value for the fines, breccias and igneous rocks (i.e. the higher the total carbon the higher the $\delta^{13}C$ value). The true situation may represent a combination of all three mechanisms.

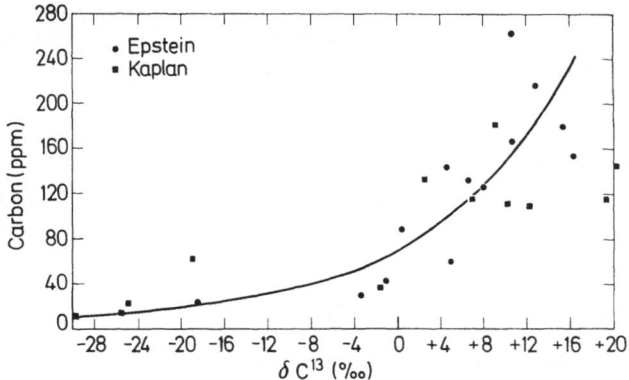

Fig. 10. Relationship between total carbon and $\delta^{13}C$ value for a variety of fines, breccias and igneous rocks from all of the Apollo missions

$\delta^{13}C$ measurements of the liquid nitrogen-fractionated gases released during stepwise pyrolysis have shown that the total carbon present is not a single homogeneous phase[48, 63]. At present insufficient information is available to identify the different phases and more detailed isotope fractionation studies are necessary.

The values of the acid (HCl) hydrolysis products of one sample of Apollo 12 fines have been measured $(+14\,^0/oo)$[48]. This value is different from those observed for known meteoritic carbide ($-4$ to $-8\,^0/oo$), suggesting that lunar carbide is distinct from meteoritic carbide. This is consistent with the origins proposed above for the major part of lunar carbide. Progressive etch of a sample of Apollo 14 fines with $H_2SO_4$ has shown that the species released become progressively enriched in $^{13}C$ with depth of etch[110]. This observation is not readily explicable in terms of any of the isotopic fractionation mechanisms proposed except preferential condensation[47].

$^{14}C$ has been shown to decrease in going from the 0–5 mm to the 20–65 mm depth of an Apollo 12 igneous rock. It has been suggested that this originates in part from cosmic ray bombardment and in part from direct implantation of $^{14}C$ from the solar wind[112].

## XII. Carbon Chemistry as an Exposure and Reworking Parameter

The concentrations of $CH_4$ and $CD_4$ released by acid dissolution correlate with parameters indicative of lunar surface exposure. As a corollary, the carbon chemistry of the regolith is also an indicator of exposure and reworking and should contribute as such to an understanding of the history of the regolith. In general, samples from the Apollo 11, 14 and 15 sites suggest a mature, extensively reworked regolith (high $CH_4$ and carbide concentrations). The Apollo

12 site at Oceanus Procellarum is less mature, (lower $CH_4$ and carbide), consistent with its location on a ray of ejecta from Copernicus. Measurements from different depths of the double core provided preliminary information about the formation and reworking of the regolith at this site[47].

The breccias returned from the Apollo 14 mission were classified[6, 113] into three basic types from visual examination, *viz*:

1. Those comprising consolidated mare material (porous and unshocked, F—1) *i.e.* typical regolith breccias,

2. those shock compressed (F—2), and

3. those thermally metamorphosed (F—3, F—4).

A simple classification based on the $CH_4$ and carbide contents parallels this classification. The F—2, F—3 and F—4 breccias contain low concentrations of both species, in agreement with the hypothesis[114] that these clastic rocks were transported to the Fra Mauro site at the time of the impact which produced the Imbrium Basin. Either the material was ejected from considerable depth, *i. e.* not exposed on the surface, or the gaseous hydrocarbons and a high proportion of the carbide (as gaseous species) were lost at the temperature attained by the ejecta[47, 115].

## XIII. Analytical Artifacts

The problems of excluding contamination from samples with low concentrations of carbon are well recognized. Analyses of lunar samples have also been troubled by another type of contaminant — artifacts — species synthesized in the samples after collection, especially during analysis. For example, variations in the quantity of aromatic hydrocarbons released are evident when the fines are pyrolysed by different methods (see above). During the analysis of lunar fines it was found that:

1. Hydrocarbons can be generated by *in vacuo* crushing in a stainless steel ball mill[45].

2. CO can be synthesised by the reaction of carbon compounds and oxygen in the ion source of a mass spectrometer[21].

3. Organosiloxanes from unknown sources can be introduced into the samples during analysis[116].

4. CO can be liberated from pyrex glass by mineral acid attack at 200 °C during acid dissolution experiments[47].

It has been suggested[24] that the $CH_4$ released by deuterated acid dissolution could arise from reaction of lunar carbide on grain surfaces with adsorbed terrestrial $H_2O$. This has been shown to be unlikely; exposure of the fines to $D_2O$ at 200 °C[85], or at ambient for periods up to several weeks[47], failed to release $CD_4$. Also, dissolution of freshly exposed chips of an Apollo 11 regolith breccia released $CH_4$ and $CD_4$ in concentrations similar to those released from Apollo 11 fines[47].

Analytical artifacts can, however, be useful when the reactions giving rise to them are recognised; for example, detection of carbide by deuterated acid hydrolysis (see above).

## XIV. Compounds of N, P and S

The lunar chemistry of nitrogen, phosphorus and sulphur has been studied in less detail than that of carbon, and less information is available about the nature and distribution of the compounds of these elements. The available data, including stable isotopic distributions, have been reviewed by Kaplan[64] and will not be discussed in detail here.

In general, total nitrogen abundances follow the same trend as total carbon abundances, although contamination by adsorbed terrestrial nitrogen remains a problem [54-56]. Thus, the igneous rocks are depleted in comparison with the fines; this suggests that a significant proportion of the nitrogen in the fines could derive from solar wind implantation, the estimated abundance of nitrogen in the solar wind being similar to that of carbon[70]. This is in agreement with the finding of Müller[44] that the concentration of chemically-bound (hydrolysable) nitrogen in a sample of Apollo 15 fines increases with decreasing grain size and that hydrolysable nitrogen compounds, and not molecular nitrogen, is the major component present. Nitrides and $NH_3$, by analogy with the presence of carbide and methane in the fines, might be expected to be synthesised in the fines and could be the species observed by acid hydrolysis[90].

Little can be said at present about phosphorus in the samples. There appears to be no systematic variation between the concentrations in the igneous rocks and the fines[64]. Deuterated acid hydrolysis of a number of samples of fines released $PD_3$, indicating the presence of phosphides[43].

Sulphur, mainly as troilite (FeS), appears to be present in higher concentrations in the igenous rocks than in the fines and breccias[56,64]. At present it is not known if this depletion in the fines and breccias is a result of losses during or after the formation of the latter from the rocks or of a mixing in of low sulphur material, or both[56,64]. The fines show an enrichment in the heavy isotope $^{34}S$ and it has been suggested that the "hydrogen stripping" process described above for carbon isotope fractionation, is responsible[64]. Sulphur could, therefore, have been removed from the fines to some extent in this way.

## XV. Conclusions

Complex organic molecules, including those of biological interest, are virtually absent from the lunar surface. The species giving rise to simple amino acids on acid hydrolysis require direct investigation. A similar situation applies to possible precursors of the porphyrin-like species which may be present in the lunar fines.

The available data show that the carbon chemistry of the lunar regolith is highly dependent on the processes occuring at the lunar surface, including solar

G. Eglinton, J. R. Maxwell and C. T. Pillinger

wind implantation and meteorite impact. There is a small magmatic component of methane in some crystalline rocks but the gaseous hydrocarbons in the fines derive mainly from solar wind implantation, although the detailed mechanisms of synthesis, and the possible stable isotope fractionation effects, are not known. The carbide in the fines derives from solar wind implantation, directly from meteorites (and possibly also by an indirect mechanism) although the relative contributions from these sources are unknown. The major part of the carbon in the fines is uncharacterised at present. The carbon accounted for as carbide in the form of $CD_4$ from the deuterated acid dissolution method represents a minimum. However, the carbide could still account for a major proportion of lunar carbon, but the precise nature of the carbides, and therefore the yield of gaseous products on acid dissolution, remains uncertain. Laboratory simulation studies offer one method of elucidating the origin and nature of the carbide.

The carbon chemistry of the regolith has been established as a significant indicator of exposure and reworking. In conjunction with other parameters indicative of exposure of the fines and breccias on the lunar surface, it should contribute to an understanding of the complex history of the regolith.

The implantation effects occuring on the lunar surface also have general implications for the synthesis of the molecules detected in interplanetary and interstellar dust [117,118]

*Acknowledgments.* We thank the Science Research Council for financial assistance and the award of a Fellowship (C.T.P.). We are grateful to the following for communicating unpublished data: Professor E. Anders, University of Chicago; Dr. N. Bhandari, Tata Institute of Fundamental Research, Bombay; Dr. D. D. Bogard, Manned Spacecraft Center, Houston; Professor K. Marti, University of California, San Diego; Dr. M. Maurette, Centre de Spectrométrie de Masse du C.N.R.S. Orsay; Professor R. O. Pepin, University of Minnesota, Minneapolis; Professor P. Signer, Swiss Federal Institute of Technology, Zurich. We thank Professor S. W. Fox, University of Miami, and Dr. E. K. Gibson, Manned Spacecraft Center, Houston, for providing figures 1 and 3 respectively. We are particularly grateful to Dr. P. H. Cadogan and Mr. B. J. Mays for their efforts in the research programme at Bristol.

## XVI. References

1) Turkevich, A. L., Franzgrote, E. J., Patterson, J. H.: Science *158,* 635 (1967).
2) Turkevich, A. L., Franzgrote, E. J., Patterson, J. H.: Science *160,* 1108 (1968).
3) Turkevich, A. L., Franzgrote, E. J., Patterson, J. H.: Science *162,* 117 (1968).
4) LSPET (Lunar Sample Preliminary Examination Team): Science *165,* 1211 (1969).
5) LSPET (Lunar Sample Preliminary Examination Team): Science *167,* 1325 (1970).
6) LSPET (Lunar Sample Preliminary Examination Team): Science *173,* 681 (1971).
7) LSPET (Lunar Sample Preliminary Examination Team): Science *175,* 407 (1972).
8) Draffan, G. H., Eglinton, G., Hayes, J. M., Maxwell, J. R., Pillinger, C. T.: Chem. Brit. *5,* 296 (1969).
9) Oyama, V. I., Merek, E. L., Silverman, M. P.: Proceedings of the Apollo 11 Lunar Science Conference (ed. A. A. Levinson), Vol. 2, pp. 1921–1927. Pergamon 1970.
10) Oyama, V. I., Merek, E. L., Silverman, M. P., Boylen, C. W.: Proceedings of the Second Lunar Science Conference (ed. A. A. Levinson), Vol. 2, pp. 1931–1937. M.I.T. Press 1971.

11) Taylor, G. R., Johnson, P. H., Kropp, K., Groves, T.: Proceedings of the Second Lunar Science Conference (ed. A. A. Levinson), Vol. 2, pp. 1939–1948, M.I.T. Press 1971.

12) Mitchell, F. J., Ellis, W. L.: Proceedings of the Second Lunar Science Conference (ed. A. A. Levinson), Vol 3, pp. 2721–2733. M.I.T. Press 1971.

13) Barghoorn, E. S.: Proceedings of the Apollo 11 Lunar Science Conference (ed. A. A. Levinson), Vol. 2, pp. 1775–1777. Pergamon 1970.

14) Schopf, J. W.: Proceedings of the Apollo 11 Lunar Science Conference (ed. A. A. Levinson), Vol. 2, pp. 1933–1934. Pergamon 1970.

15) Schopf, J. W.: Proceedings of the Second Lunar Science Conference (ed. A. A. Levinson), Vol. 2, pp. 1929–1930. M.I.T. Press 1971.

16) Cloud, P., Margolis, S. V., Moorman, M., Barker, J. M. Licari, G. R.: Proceedings of the Apollo 11 Lunar Science Conference (ed. A. A. Levinson), Vol. 2, pp. 1793–1798. Pergamon 1970.

17) Abell, P. I., Draffan, G. H., Eglinton, G., Hayes, J. M., Maxwell, J. R., Pillinger, C. T.: Proceedings of the Apollo 11 Lunar Science Conference (ed. A. A. Levinson), Vol. 2, pp. 1757–1773. Pergamon 1970.

18) Kvenvolden, K. A., Chang, S., Smith, J. W., Flores, J., Pering, K., Saxinger, C., Woeller, F., Keil, K., Breger, I. A., Ponnamperuma, C.: Proceedings of the Apollo 11 Lunar Science Conference (ed. A. A. Levinson), Vol. 2, pp. 1813–1828. Pergamon 1970.

19) Eglinton, G., Maxwell, J. R., Pillinger, C. T.: Space Life Sci. 3, 497 (1972).

20) Burlingame, A. L., Calvin, M., Han, J., Henderson, W., Reed, W., Simoneit, B. R.: Proceedings of the Apollo 11 Lunar Science Conference (ed. A. A. Levinson), Vol. 2, pp. 1779–1791. Pergamon 1970.

21) Henderson, W., Kray, W. C., Newman, W. A., Reed, W. E., Simoneit, B. R., Calvin, M.: Proceedings of the Second Lunar Science Conference (ed. A. A. Levinson), Vol. 2, pp. 1901–1912. M.I.T. Press 1971.

22) Murphy, R. C., Preti, G., Nafissi-V, M. M., Biemann, K.:Proceedings of the Apollo 11 Lunar Science Conference (ed. A. A. Levinson), Vol. 2, pp. 1891–1900. Pergamon 1970.

23) Oro, J., Updegrove, W. S., Gibert, J., McReynolds, J., Gil-Av, E., Ibanez, J., Zlatkis, A., Flory, D. A., Levy, R. L., Wolf, C. J.: Proceedings of the Apollo 11 Lunar Science Conference (ed. A. A. Levinson), Vol. 2, pp. 1901–1920. Pergamon 1970.

24) Oro, J. Flory, D. A., Gibert, J. M., McReynolds, J., Lichtenstein, H. A., Wikstrom, S.: Proceedings of the Second Lunar Science Conference (ed. A. A. Levinson), Vol. 2, pp. 1913–1925. M. I. T. Press 1971.

25) Murphy, M. E. Modzeleski, V. E., Nagy, B., Scott, W. M., Young, M., Drew, C. M., Hamilton, P. B., Urey, H. C.: Proceedings of the Apollo 11 Lunar Science Conference (ed. A. A. Levinson), Vol. 2, pp. 1879–1890. Pergamon 1970.

26) Nagy, B., Modzeleski, J. E., Modzeleski, V. E., Jabba Mohammed, M. A., Nagy, L. A., Scott, W. M., Drew, C. M., Thomas, J. E., Ward, R., Hamilton, P. B., Urey, H. C.: Nature 232, 94 (1971).

27) Rho, J. H., Bauman, A. J., Yen, T. F., Bonner, J.: Proceedings of the Apollo 11 Lunar Science Conference (ed. A. A. Levinson), Vol. 2, pp. 1929–1932. Pergamon 1970.

28) Rho, J. H., Bauman, A. J., Yen, T. F., Bonner, J.: Proceedings of the Second Lunar Science Conference (ed. A. A. Levinson), Vol. 2, pp. 1875–1877. M.I.T. Press 1971.

29) Rho, J. H., Bauman, A. J., Cohen, E. A., Yen, T. F., Bonner, J.: Space Life Sci. 3, 415 (1972).

30) Hodgson, G. W., Bunnenberg, E., Halpern, B., Peterson, E., Kvenvolden, K. A., Ponnamperuma, C.: Proceedings of the Apollo 11 Lunar Science Conference (ed. A. A. Levinson), Vol. 2, pp. 1829–1844. Pergamon 1970.

31) Hodgson, G. W., Bunnenberg, E., Halpern, B., Peterson, E., Kvenvolden, K. A., Ponnamperuma, C.: Proceedings of the Second Lunar Science Conference (ed. A. A. Levinson), Vol. 2, pp. 1865–1874. M.I.T. Press 1971.

32) Hodgson, G. W., Kvenvolden, K. A., Peterson, E., Ponnamperuma, C.: Space Life Sci. 3, 419 (1972).

109

33) Meinschein, W. G., Jackson, T. J., Mitchell, J. M., Cordes, E., Shiner, V. J.: Proceedings of the Apollo 11 Lunar Science Conference (ed. A. A. Levinson), Vol. 2, pp. 1875–1877. Pergamon 1970.

34) Mitchell, J. M., Jackson, T. J., Newlin, R. P., Meinschein, W. G.: Proceedings of the Second Lunar Science Conference (ed. A. A. Levinson), Vol. 2, pp. 1927–1928. M.I.T. Press 1971.

35) Lipsky, S. R., Cushley, R. J., Horvath, C. G., McMurray, W. J.: Proceedings of the Apollo 11 Lunar Science Conference (ed. A. A. Levinson), Vol. 2, pp. 1871–1873. Pergamon 1970.

36) Chang, S., Kvenvolden, K. A.: Exobiology (ed. C. Ponnamperuma). North Holland Publishing Company, in press.

37) Hare, P. E., Harada, K., Fox, S. W.: Proceedings of the Apollo 11 Lunar Science Conference (ed. A. A. Levinson), Vol. 2, pp. 1799–1803. Pergamon 1970.

38) Harada, K., Hare, P. E., Windsor, C. R., Fox, S. W.: Science *173*, 433 (1971).

39) Fox, S. W., Harada, K., Hare, P. E.: Space Life Sci. *3*, 425 (1972).

40) Gehrke, C. W., Zumwalt, R. W., Stalling, D. L., Roach, D., Aue, W. A., Ponnamperuma, C., Kvenvolden, K. A.: J. Chromatog. *59*, 305 (1971).

41) Gehrke, C. W., Zumwalt, R. W., Kuo, K., Aue, W. A., Stalling, D. L., Kvenvolden, K. A., Ponnamperuma, C.: Space Life Sci. *3*, 439 (1972).

42) Biemann, K.: Space Life Sci. *3*, 469 (1972).

43) Holland, P. T., Simoneit, B. R., Wszolek, P. C., Burlingame, A. L.: Proceedings of the Third Lunar Science Conference (ed. D. Heymann), Vol. 2, pp. 2131–2147. M.I.T. Press 1972.

44) Müller, O.: Lunar Science – III (ed. C. Watkins), pp. 568–570. Lunar Science Institute 1972.

45) Abell, P. I., Eglinton, G., Maxwell, J. R., Pillinger, C. T., Hayes, J. M.: Nature *226*, 251 (1970).

46) Cadogan, P. H., Eglinton, G., Maxwell, J. R., Pillinger, C. T.: Nature *231*, 29 (1971).

47) Cadogan, P. H., Eglinton, G., Firth, J. N. M., Maxwell, J. R., Mays, B. J., Pillinger, C. T.: Proceedings of the Third Lunar Science Conference (ed. D. Heymann), Vol. 2, pp. 2069–2090. M.I.T. Press 1972.

48) Chang, S., Kvenvolden, K. A., Lawless, J., Ponnamperuma, C., Kaplan, I. R.: Science *173*, 474 (1971).

49) Holland, P. T., Simoneit, B. R., Wszolek, P. C., McFadden, W. H., Burlingame, A. L.: Nature *235*, 106 (1972).

50) Holland, P. T., Simoneit, B. R., Wszolek, P. C., Burlingame, A. L.: Space Life Sci. *3*, 551 (1972).

51) Pillinger, C. T., Cadogan, P. H., Eglinton, G., Maxwell, J. R., Mays, B. J., Grant, W. A., Nobes, M. J.: Nature *235*, 108 (1972).

52) Eglinton, G., Maxwell, J. R., Pillinger, C. T.: Sci. Am. *227*, 80 (1972).

53) Lawless, J. G., Folsome, C. E., Kvenvolden, K. A.: Sci. Am. *226*, 38 (1972).

54) Moore, C. B., Gibson, E. K., Larimer, J. W., Lewis, C. F., Nichiporuk, W.: Proceedings of the Apollo 11 Lunar Science Conference (ed. A. A. Levinson), Vol. 2, pp. 1375–1382. Pergamon 1970.

55) Moore, C. B., Lewis, C. F., Larimer, J. W., Delles, F. M., Gooley, R. C., Nichiporuk, W., Gibson, E. K.: Proceedings of the Second Lunar Science Conference (ed. A. A. Levinson), Vol. 2, pp. 1343–1350. M.I.T. Press 1971.

56) Moore, C. B., Lewis, C. F., Cripe, J., Delles, F. M., Kelly, W. R., Gibson, E. K.: Proceedings of the Third Lunar Science Conference (ed. D. Heymann), Vol. *2*, pp. 2051–2058. M.I.T. Press 1972.

57) Epstein, S., Taylor, H. P.: Proceedings of the Apollo 11 Lunar Science Conference (ed. A. A. Levinson), Vol. 2, pp. 1085–1096. Pergamon 1970.

58) Epstein, S., Taylor, H. P.: Proceedings of the Apollo 12 Lunar Science Conference (ed. A. A. Levinson), Vol. 2, pp. 1421–1441. M.I.T. Press 1971.

59) Epstein, S., Taylor, H. P.: Proceedings of the Third Lunar Science Conference (ed. D. Heymann), Vol. 2, pp. 1429–1454. M.I.T. Press 1972.

60) Friedman, I., Gleason, J. D., Hardcastle, K. G.: Proceedings of the Apollo 11 Lunar Science Conference (ed. A. A. Levinson), Vol. 2, pp. 1103–1109. Pergamon 1970.

61) Friedman, I., O'Neill, J. R., Gleason, J. D., Hardcastle, K.: Proceedings of the Second Lunar Science Conference (ed. A. A. Levinson), Vol. 2, pp. 1407–1415. M.I.T. Press 1971.

62) Kaplan, I. R., Smith, J. W., Ruth, E.: Proceedings of the Apollo 11 Lunar Science Conference (ed. A. A. Levinson), Vol. 2, pp. 1317–1329. Pergamon 1970.

63) Kaplan, I. R., Petrowski, C.: Proceedings of the Apollo 12 Lunar Science Conference (ed. A. A. Levinson), Vol. 2, pp. 1397–1406. M.I.T. Press 1971.

64) Kaplan, I. R.: Space Life Sci. 3, 383 (1972).

65) Mason, B.: Meteorites, Chap. 5. Wiley & Sons 1962.

66) Gonapathy, R., Keays, R. R., Laul, J. C., Anders, E.: Proceedings of the Apollo 11 Lunar Science Conference (ed. A. A. Levinson), Vol. 2, pp. 1117–1142. Pergamon 1970.

67) Laul, J. C., Morgan, J. W., Ganapathy, R., Anders, E.: Proceedings of the Second Lunar Science Conference (ed. A. A. Levinson), Vol. 2, pp. 1139–1158. M.I.T. Press 1971.

68) Wasson, J. T., Baedecker, P. A.: Proceedings of the Apollo 11 Lunar Science Conference (ed. A. A. Levinson), Vol. 2, pp. 1741–1750. Pergamon 1970.

69) Neukum, G.: Personal communication, 1972.

70) Cameron, A. G. W.: Origin and distribution of the elements (ed. L. H. Ahrens), pp. 125–143. Pergamon 1968.

71) Hayes, J. M.: Space Life Sci. 3, 474 (1972).

72) Anderson, A. T., Crew, A. V., Goldsmith, J. R., Moore, P. B., Newton, J. C., Olsen, E. J., Smith, J. V., Wyllie, P. J.: Science 167, 587 (1970).

73) Frondel, C., Klein, C., Ito, J., Drake, J. E.: Proceedings of the Apollo 11 Lunar Science Conferences (ed. A. A. Levinson), Vol. 1, pp. 445–474. Pergamon 1970.

74) Adler, I., Walter, L. S., Lowman, P. D., Glass, B. P., French, B. M., Philpotts, J. A., Heinrich, K. J. F., Goldstein, J. I.: Proceedings of the Apollo 11 Lunar Science Conference (ed. A. A. Levinson), Vol. 1, pp. 87–92. Pergamon 1970.

75) Reed, G. W., Jovanovic, S., Fuchs, L. H.: Lunar Science – III (ed. C. Watkins), pp. 637–639. Lunar Science Institute 1972.

76) Arrhenius, G., Liang, S., McDougall, D., Wilkenning, L., Bhandari, N., Bhat, S., Lal, D., Rajagopalan, G., Tamhane, A. S., Venkatavaradan, V. S., Proceedings of the Second Lunar Science Conference (ed. A. A. Levinson), Vol. 3, pp. 2583–2598. M.I.T. Press 1971.

77) Gay, P., Bancroft, G. M., Bown, M. G.: Proceedings of the Apollo 11 Lunar Science Conference (ed. A. A. Levinson), Vol. 1, pp. 481–497. Pergamon 1970.

78) Flory, D. A., Simoneit, B. R.: Space Life Sci. 3, 457 (1972).

79) Burlingame, A. L., Hauser, J. S., Simoneit, B. R., Smith, D. H., Biemann, K., Mancuso, N., Murphy, R.: Proceedings of the Second Lunar Science Conference (ed. A. A. Levinson), Vol. 2, pp. 1891–1899. M.I.T. Press 1971.

80) Gibson, E. K., Moore, G.: Proceedings of the Third Lunar Science Conference (ed. D. Heymann), Vol. 2, pp. 2029–2040. M.I.T. Press 1972.

81) Gibson, E. K., Johnson, S. M.: Proceedings of the Second Lunar Science Conference (ed. A. A. Levinson), Vol. 2, pp. 1351–1366. M.I.T. Press 1971.

82) Gibson, E. K., Moore, C. B.: Space Life Sci. 3, 404 (1972).

83) Chang, S., Smith, J. W., Kaplan, I. R., Lawless, J., Kvenvolden, K. A., Ponnamperuma, C.: Proceedings of the Apollo 11 Lunar Science Conference (ed. A. A. Levinson), Vol. 2, pp. 1857–1869. Pergamon 1970.

84) Preti, G., Murphy, R. C., Biemann, K.: Proceedings of the Second Lunar Science Conference (ed. A. A. Levinson), Vol. 2, pp. 1879–1889. M.I.T. Press 1971.

85) Flory, D. A., Wikstrom, S., Gupta, S., Gibert, J. M., Oro, J.: Proceedings of the Third Lunar Science Conference (ed. D. Heymann), Vol. 2, pp. 2091–2108. M.I.T. Press 1972.

G. Eglinton, J. R. Maxwell and C. T. Pillinger

86) Abell, P. I., Cadogan, P. H., Eglinton, G., Maxwell, J. R., Pillinger, C. T.: Proceedings of the Second Lunar Science Conference (ed. A. A. Levinson), Vol. 2, pp. 1843–1863. M.I.T. Press 1971.
87) Funkhouser, J., Jessberger, E., Müller, O., Zahringer, J.: Proceedings of the Second Lunar Science Conference (ed. A. A. Levinson), Vol. 2, pp. 1381–1396. M.I.T. Press 1971.
88) Quaide, W., Oberbeck, V., Bunch, T.: Proceedings of the Second Lunar Science Conference (ed. A. A. Levinson), Vol. 1, pp. 701–718. M.I.T. Press 1971.
89) McKay, D. S., Morrison, D. A., Clanton, U. S., Ladle, G. H., Lindsay, J. F.: Proceedings of the Second Lunar Science Conference (ed. A. A. Levinson), Vol. 1, pp. 755–773. M.I.T. Press 1971.
90) Hintenberger, H., Weber, H. W., Voshage, H., Wänke, H., Begemann, F., Wlotzka, F.: Proceedings of the Apollo 11 Lunar Science Conference (ed. A. A. Levinson), Vol. 2, pp. 1269–1282. Pergamon 1970.
91) Marti, K.: Personal communication, 1971.
92) Pepin, R. O.: Personal communication, 1971.
93) Bogard, D. D.: Personal communication, 1972.
94) Funkhouser, J.: Personal communication, 1972.
95) Signer, P., Schultz, L.: Personal communication, 1972.
96) Arrhenius, G., Asunmaa, S., Drever, J. I., Everson, J. E., Fitzgerald, R. W., Frazer, J. Z., Fujita, H., Hanor, J. S., Lal, D., Liang, S. S., MacDougall, D., Reid, A., Sinkankas, J., Wilkenning, L.: Science 167, 659 (1970).
97) Phakey, P. P., Hutcheon, D., Rajan, R. S., Price, P. B.: Lunar Science – III (ed. C. Watkins), pp. 608–610. Lunar Science Institute 1972.
98) Poupeau, G., Berdot, J. L., Chetrit, G. C., Pellas, P.: Lunar Science – III (ed. C. Watkins), pp. 613–615. Lunar Science Institute 1972.
99) Bhandari, N.: Personal communication, 1972.
100) Borg, J., Maurette, M., Durrieu, L., Jouret, C.: Proceedings of the Second Lunar Science Conference (ed. A. A. Levinson), Vol. 3, pp. 2027–2040. M.I.T. Press 1971.
101) Dran, J. C., Durrieu, L., Jouret, C., Maurette, M.: Earth Planet. Sci. Letters 9, 391 (1970).
102) Bibring, J. P., Maurette, M., Meunier, R., Durrieu, L., Jouret, C., Eugster, O.: Lunar Science – III (ed. C. Watkins), pp. 71–73. Lunar Science Institute 1972.
103) Maurette, M.: Personal communication, 1972.
104) Carter, J. L., Macgregor, I. D.: Proceedings of the Apollo 11 Lunar Science Conference (ed. A. A. Levinson), Vol. 1, pp. 247–265. Pergamon 1970.
105) Goldstein, J. I., Henderson, E. P., Yakowitz, H.: Proceedings of the Apollo 11 Lunar Science Conference (ed. A. A. Levinson), Vol. 1, pp. 499–512. Pergamon 1970.
106) Housley, R. M., Grant, R. W., Abdel-Gawad, M.: Lunar Science – III (ed. C. Watkins), pp. 392–394. Lunar Science Institute 1972.
107) Carter, J. L.: Lunar Science – III (ed. C. Watkins), pp. 125–127. Lunar Science Institute 1972.
108) Morgan, J. W., Krähenbuhl, U., Ganapathy, R., Anders, E.: Lunar Science – III (ed. C. Watkins), pp. 555–557. Lunar Science Institute 1972.
109) Cadogan, P. H., Eglinton, G., Maxwell, J. R., Pillinger, C. T.: Nature 241, 81 (1973).
110) Sakai, H., Petrowski, C., Goldhaber, M. B., Kaplan, I. R.: Lunar Science – III (ed. C. Watkins), pp. 672–674. Lunar Science Institute 1972.
111) Berger, R.: Nature 226, 738 (1970).
112) Begemann, F., Born, W., Palme, E., Vilcsek, E., Wänke, H.: Lunar Science – III (ed. C. Watkins), pp. 53–54. Lunar Science Institute 1972.
113) Jackson, E. D., Wilshire, H. G.: Lunar Science – III (ed. C. Watkins), pp. 418–420. Lunar Science Institute 1972.
114) Anderson, A. T., Braziunas, T. F., Jacoby, J., Smith, J. V.: Lunar Science – III (ed. C. Watkins), pp. 24–26. Lunar Science Institute 1972.
115) Williams, R. J.: Personal communication, 1972.

116) Gehrke, C. W., Zumwalt, R. W., Aue, W. A., Stalling, D. L., Duffield, A., Kvenvolden, K. A., Ponnamperuma, C.: Proceedings of the Apollo 11 Lunar Science Conference (ed. A. A. Levinson), Vol. 2, pp. 1845–1856. Pergamon 1970.
117) Buhl, D.: Nature *234*, 332 (1971).
118) Rank, D. M., Townes, C. H., Welch, W. J.: Science *174*, 1083 (1971).

Received July 10, 1972

# Chemistry of the Moon

**Prof. Dr. Heinrich Wänke**
Max-Planck-Institut für Chemie (Otto-Hahn-Institut), Mainz

## Contents

I.  Introduction . . . . . . . . . . . . . . . . . 116
II. The Chemistry of the Lunar Samples . . . . . . . . . 117
   A. Types of Samples Returned . . . . . . . . . . . 117
   B. Mare Basalts . . . . . . . . . . . . . . . . 119
   C. Non-Mare Igneous Rocks . . . . . . . . . . . . 124
   D. The Lunar Regolith . . . . . . . . . . . . . . 124
   E. KREEP . . . . . . . . . . . . . . . . . . 126
III. The Europium-Anomaly . . . . . . . . . . . . . 129
IV.  Solar Wind Implantation . . . . . . . . . . . . 130
V.   Meteoritic Component . . . . . . . . . . . . . 133
VI.  Formation of the Lunar Landscape . . . . . . . . . 137
VII. Element Correlations . . . . . . . . . . . . . 140
VIII. Condensation Sequence of the Planetary Nebula . . . . . 143
IX.  Composition and Structure of the Moon . . . . . . . 144
   A. Chemical Composition . . . . . . . . . . . . 144
   B. Internal Structure . . . . . . . . . . . . . . 147
X.   The Origin of the Moon . . . . . . . . . . . . 149
XI.  References . . . . . . . . . . . . . . . . . 150

H. Wänke

## I. Introduction

The most dominant features of the lunar landscape are the large maria that are
easy to distinguish from the lighter continents on all lunar photographs. For a
long time most scientists have favored the idea of a lava flow origin for the
lunar maria. From analyses of the major elements derived from the unmanned
U.S. space probes, Surveyor V, VI, and VII, it was learned that the lunar surface consists of basalt[1] (Table 1).

Table 1. Results of the instrumental analysis of the lunar surface by the unmanned space
probes Surveyor V, VI, and VII via the alpha-scattering technique (Turkevich et al.[1]).
The data obtained on Apollo 11 soil sample 10084 are given for comparison[2]

|  | Mare Tranquillitatis | | Sinus Medii | Highland near Tycho |
|---|---|---|---|---|
|  | Surveyor V[1] | Apollo 11[2]<br>(Soil 10084) | Surveyor VI[3] | Surveyor VII[4] |
| O | 43 | 41.5 | 43.8 | 44.6 |
| Na | 0.45 | 0.32 | 0.6 | 0.5 |
| Mg | 2.7 | 4.8 | 4.0 | 4.2 |
| Al | 7.6 | 6.9 | 7.8 | 11.8 |
| Si | 21.7 | 19.7 | 22.9 | 21.5 |
| Ca | 10.4 | 8.1 | 9.2 | 13.1 |
| Ti | 4.5 | 4.3 | 2.1 | 0 |
| Fe | 9.4 | 12.0 | 9.6 | 4.3 |

Laboratory study of the first lunar samples brought to Earth by Apollo 11
ruled out all ideas that the Moon might be a primitive object, *i.e.* an object that
had remained rather cool after its accumulation with only minor melting processes induced by large impacts on the surface. On the contrary, it became evident from the chemical analysis that the Moon, like the Earth, is *a highly differentiated object*. On the Moon many chemical elements are strongly enriched
or depleted as compared with their abundances in carbonaceous chondrites
which apart from the most volatile elements, are believed to be most representative of solar matter. Thus it became clear that, at least in the upper 200 km,
extensive melting processes must have occurred.

The crater density in the mare regions is about one order of magnitude
lower than on the lunar continents. Most of the lunar craters are clearly of
impact origin; we therefore have to conclude that the continents predate the
maria. The lifetime of objects crossing the orbit of the Moon and the Earth is
short compared with the age of the planetary system. So the flux of impacting
bodies must have been higher in the early days of the solar system, decreasing
considerably as time went on. The difference of a factor of about ten in the
crater density of the lunar maria and continents does not correspond to a
similar difference in age. From age determinations on lunar rocks we have now
learned that the lava filling of the maria so far investigated (Tranquillitatis,
Imbrium, Oceanus Procellarum and Oceanus Fecunditatis) occurred between

116

about 3.9 and 3.3 billion years ago, while for the continents we estimate an upper limit of about 4.4 billion years.

## II. The Chemistry of the Lunar Samples

### A. Types of Samples

The lunar samples can be divided into three classes:
a) Crystalline igneous rocks. These range from very fine-grained vesicular rocks to vuggy, medium-grained equigranular rocks, in rare cases containing crystals more than 1 cm in diameter.
b) Breccias. These vary from fine-grained micro-breccias to clasts containing large fragments of igneous rocks. The breccias consist of a mechanical mixture of soil and small rock fragments compacted to a coherent rock.
c) Soil, or lunar fines. The soil is a loose mixture of crystalline grains with larger aggregates, glassy fragments (including many spherules) and trace amounts of metallic iron particles.

Impacts of extralunar objects (meteorites, asteroids, comets) play a very important role in shaping the lunar landscape (craters of all sizes, production of lunar soil) but in general extralunar matter does not make up a significant proportion of the lunar surface samples. Even in the case of the lunar soil, the meteoritic contribution amounts to only a few percent (see Chap. V). The meagerness of this contribution is explained by the fact that a meteoritic object impacting on the lunar surface at an average velocity of about 15 km/sec leaves a mass of crushed rocks about 200 times larger than the projectile mass[5]. Thus, except for the highly siderophile and highly volatile elements, the meteoritic contribution can be neglected in the case of lunar soils and breccias; for igneous rocks, it is normally considered to be zero.

With the exception of these small amounts of meteoritic matter, the lunar soil is derived from igneous rocks disrupted by impacts and gradually reduced to fine-grained dust. On all landing sites, the soil is much more abundant than rocks or breccias. Except on steep slopes, the whole Moon is covered with a layer dust of at least 5 m thick. The rock samples brought back are separate fragments embedded in the soil; once part of the underlying bedrock, they were excavated by the impact of meteorites.

The nomenclature used for the lunar samples in this paper is that of NASA. The first two digits (or, as from Apollo 16, the first digit) of the sample number refer to the flight number and hence to the landing site (see Fig. 1):

| | |
|---|---|
| 10 | Apollo 11 Mare Tranquillitatis; typical mare |
| 12 | Apollo 12 Oceanus Procellarum; typical mare |
| 14 | Apollo 14 Fra Mauro; probably Imbrium ejecta. |
| 15 | Apollo 15 Hadley Rille; on the edge of Mare Imbrium close to Montes Apenninus |
| 6 | Apollo 16 Descartes Region; typical highland. |

Fig. 1. Front side of the Moon. Landing sites of all sample-return missions of Apollo and Luna (L16 and L 20) are marked

## B. Mare Basalts

Most of the igneous rocks studied so far are mare basalts. For the chemical composition of typical examples, see samples 10044, 10057, 12018 and 15415 in Table 2 and Fig. 2. The grain size and the vesicular texture indicate that these rocks solidified near the surface.

Fig. 2. Basalt from Mare Tranquillitatis (rock 10057). Fine-grained vesicular rock

The main minerals of the mare basalts are, in order of abundance:
pyroxene $(Mg, Fe)SiO_3$ and olivine $(Mg, Fe)_2SiO_4$,
Ca-rich plagioclase (solid solution of $CaAl_2Si_2O_8$ and $NaAlSi_3O_8$),
ilmenite $(FeTiO_3)$.

The Apollo 11 rocks contain large amounts of ilmenite, as can be seen from Tables 2 and 3 (high titanium content). We have plotted the chemical composition of rock sample 12018 in Fig. 3a $vs.$ that of the carbonaceous chondrites (the most primitive of all meteorites), in Fig. 3b $vs.$ the basaltic achondrite (eucrite) Juvinas (a class of meteorites which have undergone magmatic differentiation) and in Fig. 4 $vs.$ the average composition of the Earth's

H. Wänke

Table 2. Chemical composition of individual lunar rock and soul samples. Luna 16 and Luna 20 refer to the unmanned Soviet space probes

| | Rocks | | | | Soils | | | | | | |
|---|---|---|---|---|---|---|---|---|---|---|---|
| | 10044 | 10057 | 12018 | 15415 | 10084 | 12070 | 14163 | 15601 | 60601 | Luna 16 Regolith | Luna 20 Regolith |
| Refs. | 2) | 2) | 6) | 7, 8, 9) | 2) | 6) | 12) | 12) | 13) | 14) | 15) |
| % | | | | | | | | | | | |
| O | 41.5 | 40.4 | 42.0 | 46.2 | 41.5 | 42.6 | 43.7 | 41.8 | 44.4 | 42.2 | 44.1 |
| Mg | 3.9 | 4.2 | 8.69 | 0.097 | 4.8 | 5.84 | 5.6 | 6.8 | 3.91 | 5.3 | 5.8 |
| Al | 6.3 | 4.0 | 4.22 | 18.8 | 6.9 | 6.72 | 9.6 | 5.7 | 13.9 | 8.1 | 12.1 |
| Si | 20.1 | 18.9 | 21.1 | 20.6 | 19.7 | 21.5 | 22.6 | 21.8 | 21.2 | 19.5 | 20.7 |
| Ca | 5.1 | 8.4 | 4.4 | 14.1 | 8.1 | 7.6 | 7.3 | 6.7 | 12.0 | 8.7 | 10.9 |
| Ti | 6.3 | 6.5 | 1.59 | 0.01 | 4.3 | 1.6 | 0.87 | 0.94 | 0.35 | 2.0 | 0.33 |
| Fe | 13.3 | 14.0 | 16.5 | 0.18 | 12.0 | 12.7 | 8.1 | 15.0 | 4.31 | 13.1 | 5.5 |
| ppm | | | | | | | | | | | |
| F | 83 | 88 | – | – | 86 | 60 | 145 | 45 | 59 | 265 | 37 |
| Na | 3580 | 3000 | 1430 | 2520 | 3150 | 3090 | 5000 | 2260 | 3250 | 2750 | 4080 |
| K | 860 | 2010 | 410 | 127 | 1090 | 1900 | 4430 | 870 | 890 | 830 | 830 |
| Cr | 1300 | 2160 | 3730 | – | 1830 | 2270 | 1290 | 3540 | 720 | 2100 | 700 |
| Mn | 2000 | 1800 | 1980 | – | 1560 | 1590 | 1010 | 1880 | 540 | 1630 | 930 |
| La | 12 | 25 | 5.68 | 0.12 | 15 | 33.0 | 68 | 12.9 | 13.4 | 7.3 | 4.9 |
| Eu | 2.69 | 1.80 | 0.84 | 0.81 | 1.67 | 1.80 | 2.45 | 1.03 | 1.29 | 1.6 | 0.9 |
| W | 0.24 | 0.43 | 0.15 | – | 0.22 | 0.74 | 1.95 | 0.28 | 0.3 | – | – |
| Au | – | 0.000017[1] | – | 0.00012[1] | 0.0021 | 0.0018 | 0.0061 | 0.0016 | 0.010 | 0.0033 | – |
| U | 0.28 | 0.80 | 0.252 | 0.011 | 0.35 | 1.69 | 4.07 | 0.57 | 0.57 | 0.47 | ~0.5 |

[1]) Data for Au are taken from Anders et al.[10, 11].

Table 3. Average composition of lunar rocks. Data taken from the compilation of Schonfeld and Meyer[16]

| | Mare basalts | | | | | | | | Anorthosite clan | |
| | Apollo 11 | | Apollo 12 | | Apollo 14 | Apollo 15 | Luna 16 | KREEP | Anorthosite | Gabbroic anorthosite |
| % | low K | high K | low Mg | high Mg | | | | | | |
|---|---|---|---|---|---|---|---|---|---|---|
| Mg | 4.30 | 4.48 | 4.40 | 7.80 | 5.10 | 5.7 | 4.25 | 4.82 | 1.33 | 3.80 |
| Al | 5.48 | 4.30 | 5.50 | 6.73 | 6.80 | 4.74 | 7.23 | 9.26 | 17.73 | 14.8 |
| Si | 18.9 | 19.37 | 21.80 | 20.70 | 21.60 | 21.50 | 20.50 | 22.50 | 20.70 | 20.90 |
| Ca | 8.49 | 7.76 | 8.03 | 6.20 | 7.95 | 7.3 | 7.43 | 7.87 | 12.7 | 11.6 |
| Ti | 6.43 | 7.00 | 2.20 | 1.58 | 1.76 | 1.28 | 2.94 | 1.14 | 0.10 | 0.18 |
| Fe | 14.71 | 14.79 | 15.50 | 16.35 | 13.30 | 17.6 | 15.03 | 8.33 | 1.3 | 3.1 |
| ppm | | | | | | | | | | |
| Na | 2957 | 3580 | 1560 | 2080 | 3260 | 1930 | 2920 | 6520 | 5200 | 2600 |
| P | 440 | 750 | 400 | 400 | 570 | 310 | 480 | 4000 | – | 100 |
| K | 600 | 2500 | 510 | 450 | 912 | 380 | 1200 | 5200 | 150 | 400 |
| Cr | 1915 | 2394 | 2630 | 4210 | 2530 | 4600 | 1920 | 1300 | 400 | 200 |
| Mn | 2140 | 1843 | 2090 | 2120 | 2020 | 2250 | 1550 | 1300 | – | 200 |
| Rb | 0.812 | 5.65 | 1.1 | 0.9 | 2.1 | 0.6 | 1.9 | 16.3 | 0.32 | 1.27 |
| Sr | 170 | 167 | 115 | 90 | 101 | 98 | 370 | 200 | 160 | 160 |
| Zr | 310 | 410 | 140 | 100 | 310 | 88 | 296 | 1300 | – | 150 |
| Ba | 93 | 293 | 76 | 65 | 146 | 33 | 222 | 1030 | 30 | 61 |
| La | 12.2 | 28.1 | 5.55 | 7.35 | 13 | – | 7.7 | 93 | 1.58 | (7) |
| Eu | 2.02 | 2.24 | 1.28 | 0.91 | 1.21 | 0.7 | 2.5 | 3.25 | 0.75 | 0.75 |
| Th | 0.97 | 3.30 | 1.10 | 0.81 | 2.10 | 0.50 | – | 17.2 | – | – |
| U | 0.25 | 0.86 | 0.27 | 0.23 | 0.59 | 0.13 | – | 4.50 | 0.058 | – |

crust. To the best of our knowledge, the carbonaceous chondrites are most representative of the non-volatile solar matter.

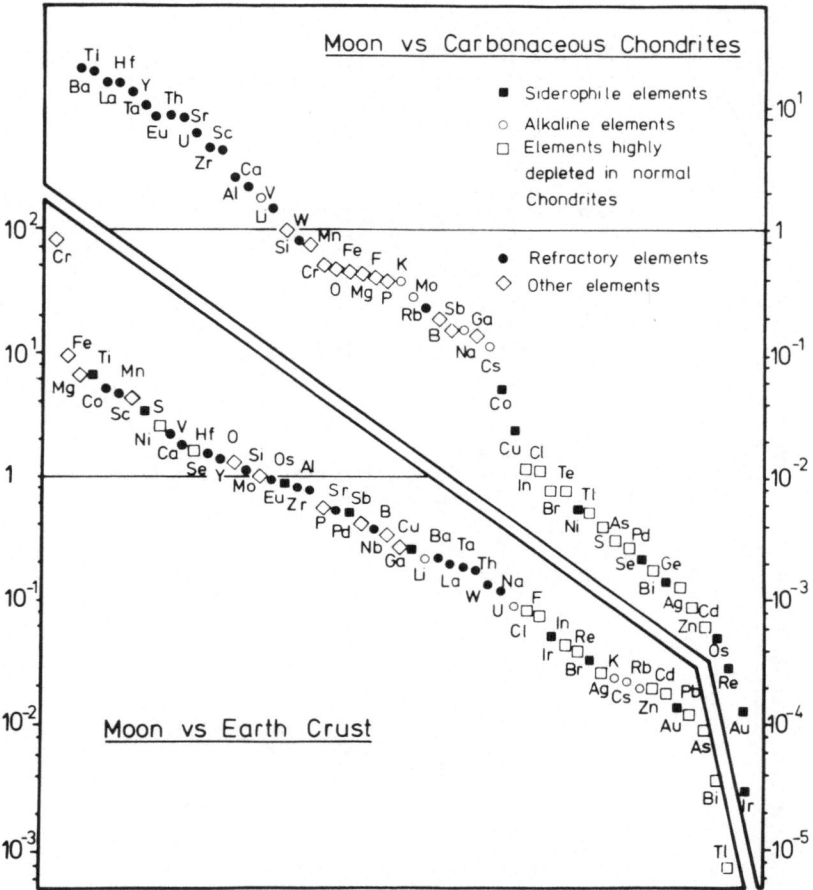

Fig. 3a. (Upper diagram). Abundances of chemical elements in lunar rock 12018 *vs.* the abundances in carbonaceous chondrites type 1 (C 1) normalized to silicon. Data for 12018 were taken from all authors of the Second Lunar Science Conference. Geochim. Cosmochim. Acta, Supplement 2, *2* (1971), data for C 1 from the compilation of Mason[17)]

Fig. 3b. (Lower diagram). Abundances of chemical elements in lunar rock 12018 *vs.* the abundances in the Earth's crust[18, 19)]

The elemental abundance of the lunar mare rocks as compared to that of carbonaceous chondrites vary up to 6 orders of magnitude (Fig. 3a). The strongly siderophile elements and the very volatile elements are highly depleted, while the refractory elements Al, Ca, Ti, REE, Th, U. etc. are enriched. Hence, it is rather difficult to explain the fractionation of the lunar mare basalts by

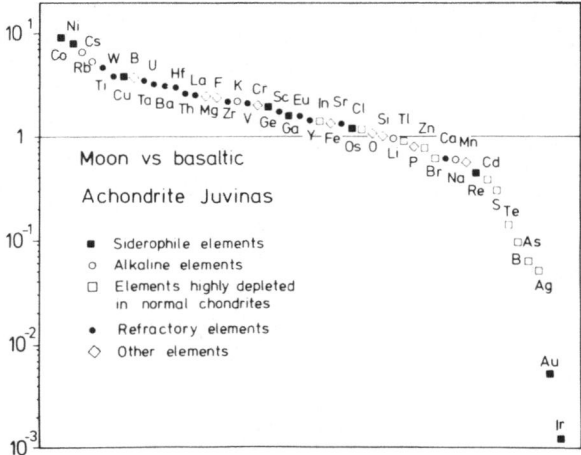

Fig. 4. Abundances of chemical elements in lunar rock 12018 *vs.* the abundances in the basaltic achondrite (eucrite) Juvinas. Data for Juvinas from Wänke *et al.*[12] and Mason[17]

magmatic processes alone. Magmatic differentiations obviously leave the K/U ratio nearly unchanged. The Earth's crust for example is enriched, not only in elements like Ca, Al, La, U, etc., but also in K. All the more volatile elements are rarer on the Moon than in the Earth's crust.

The relatively high abundance in the lunar mare rocks of elements which on Earth are concentrated in the mantle[19] – Fe, Mg, and especially Cr (Fig. 3b) could indicate that the differentiation between "crust" and "mantle" was less complete on the Moon than on Earth. Alternatively, if such a differentiation did take place on the Moon, the mare rocks must represent mantle material, *i.e.* they would have come from a considerable depth. The latter seems more likely. According to the results of the analyses of the Apollo 16 samples (see soil 60601 in Table 2), the elements Fe, Mg, and Cr are indeed much rarer in the lunar highlands, which are thought to represent the ancient lunar crust. The basaltic achondrites (eucrites as well as howardites) show the greatest similarity to the lunar mare basalts (Fig. 4). We will come back to this relation later on.

Referring to the alkali metal elements, we notice that the most refractory element, lithium, shows no depletion (Fig. 3a), as pointed out by Gast *et al.*[20] Further, depletion is not mass-dependent, which is additional evidence for the separation of volatiles from the refractory elements in the absence of a strong gravitational field.[20]

There cannot be any doubt that the lunar maria were formed by large lava flows[21]. The basalts from the various landing sites differ in composition (Table 3), but samples from a single mare site are not uniform either. Fig. 5 plots the concentrations of Al *vs.* Mg in lunar basalts from Apollo 12. The observed trend can be accounted for by magmatic differentiation processes, with the variations in chemical composition reflecting origin at different depths.

This fact, taken together with the layers seen in photographs[22] of the slopes of Hadley Rille, leads us to conclude that the maria were not filled in one step but in successive eruptions, each forming layers only tens of meters thick.

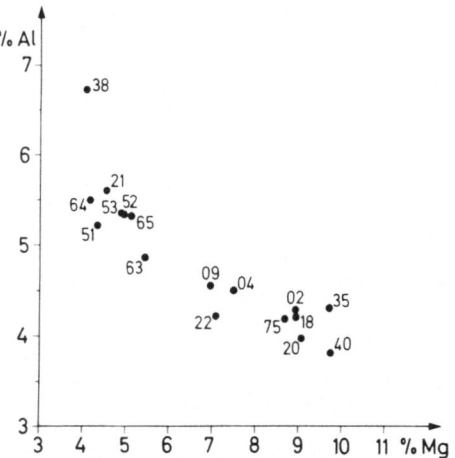

Fig. 5. Al *vs.* Mg in lunar igneous rocks from Apollo 12. From Wänke *et al.*[6]

Apollo 11 brought back two chemically differing groups of basalt: high-K and low-K rocks. From age determinations we know that these two rock types solidified more than 200 million years apart. Turner[23] found a K-Ar age of $3.55 \times 10^9$ yrs for the high-K rocks and $3.82 \times 10^9$ yrs for the low-K rocks.

## C. Non-Mare Igneous Rocks

Among the Apollo 11 samples millimeter sized grains were found, belonging to a rock type that differed in composition from the mare basalts[24]; the first large sample was detected by the Apollo 15 crew near the Apennine Mountains. This rock type consists of nearly 100% pure anorthite (calcic plagioclase) and is called anorthosite. In their paper on the discovery of anorthosites in lunar samples, Wood *et al.*[24] suggested that this material is the dominant sur-face rock of the lunar highlands. This expectation was confirmed by the samples of anorthosites brought back by the Apollo 16 mission[25], together with sam-ples of a composition intermediate between anorthosite, mare basalts and KREEP[25]. (See Chap. II. E).

## D. The Lunar Regolith

In some respects the dust layer several meters thick which covers the whole surface of the Moon gives a better idea of the surface composition of the various landing sites than do individual rock samples. The total number of rock· samples is naturally very limited and, as we have seen, the rocks from a single site are by no means compositionally homogeneous. Hence, the small grains of the lunar soils give a better average. Due to large impacts, mixing within the lunar regolith occurs over a wide area, but except for special cases the contribution of material from distances more than 100 km away is nevertheless small, as may be seen from Fig. 6.

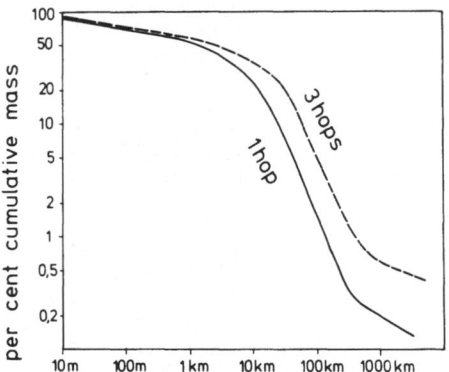

Fig. 6. Dispersion of regolith material by cratering. According to Shoemaker et al.[26]

Table 2 gives the composition of a representative soil sample from each landing site. The composition of the soils from the various locations can also be compared with the results of the chemical analyses of the lunar surface made by the unmanned space probes, Surveyor V, VI, and VII (Table 1), and the data from the x-ray fluorescence experiment (Table 4). The latter experiment made use of the fluorescence x-rays emitted from the lunar surface due to excitation by the solar x-rays which were measured on the Command Service Module of Apollo 15 during its orbit around the Moon. In this way values for the Al/Si and Mg/Si ratios were obtained for large areas of the lunar surface.[27]

Table 4 summarizes the Al/Si ratios obtained by three methods; they are in agreement within the experimental errors. The soil of the Descartes Region (Apollo 16), the only typical highland area visited by sample-return missions, has an Al/Si ratio of 0.6 (soil sample 60601), which is the highest found in all laboratory measurements of lunar soil samples and among the highest determined in the x-ray fluorescence experiment. For pure anorthositic rocks the Al/Si ratio is within the range 0.85 to 0.91. Hence, the highlands cannot consist solely of anorthosites. For the highlands distant from the large maria, contamination with mare material is unlikely.

The values for the Al/Si ratio of the regolith in mare areas, as determined from moon samples as well as via the x-ray fluorescence technique, also differ from the values found in the rock samples of mare basalts (Tables 2 and 3), the latter being somewhat lower. These differences can be explained by the admixture of anorthositic material from the lunar highlands, as found at the

125

Table 4. Comparison of x-ray fluorescence experiment data (XRFE)[27] with those from returned-sample analysis (RSA) and the alpha-backscatter technique (AST). See Table 1 and 2

| Location | Al/Si ± 1σ | Mg/Si ± 1σ | Method |
|---|---|---|---|
| South part of Gagarin 144° − 153° E, 21 − 23° S | 0.65 ± 0.24 | 0.14 ± 0.05 | XRFE |
| Highland around Tsiolkovsky | 0.62 ± 0.12 | 0.15 ± 0.06 | XRFE |
| Tsiolkovsky Rim | 0.54 ± 0.12 | 0.16 ± 0.02 | XRFE |
| Tsiolkovsky | 0.39 ± 0.11 | 0.18 ± 0.02 | XRFE |
| Highland east of Smythii | 0.62 ± 0.07 | 0.19 ± 0.04 | XRFE |
| Mare Smythii | 0.45 ± 0.06 | 0.27 ± 0.06 | XRFE |
| Highland south of Crisium | 0.53 ± 0.06 | 0.23 ± 0.03 | XRFE |
| Luna 20 regolith | 0.58 | 0.28 | RSA |
| Mare Crisium | 0.39 ± 0.08 | 0.26 ± 0.05 | XRFE |
| Mare Fecunditatis | 0.36 ± 0.06 | 0.25 ± 0.03 | XRFE |
| Luna 16 regolith A | 0.42 | 0.28 | RSA |
| Mare Tranquillitatis | 0.34 ± 0.06 | 0.25 ± 0.04 | XRFE |
| Soil 10084 (Tranquillitatis) | 0.35 | 0.24 | RSA |
| Mare Serenitatis | 0.28 ± 0.08 | 0.21 ± 0.06 | XRFE |
| Haemus Mountains | 0.38 ± 0.10 | 0.25 ± 0.05 | XRFE |
| Apennine Mountains | 0.36 ± 0.06 | 0.25 ± 0.03 | XRFE |
| Palus Putredinus | 0.35 ± 0.09 | 0.30 ± 0.03 | XRFE |
| Soil 15601 (Mare Imbrium) | 0.26[1]) | 0.31 | RSA |
| Mare Imbrium | 0.30 ± 0.10 | 0.21 ± 0.06 | XRFE |
| North-east of Schröters Valley | 0.26 ± 0.13 | 0.21 ± 0.06 | XRFE |
| Soil 12070 (Oceanus Procellarum) | 0.31 | 0.27 | RSA |
| Soil 14163 (Fra Mauro; KREEP rich) | 0.42 | 0.25 | RSA |
| Soil 60601 (Descartes; anorthosite − rich) | 0.66 | 0.18 | RSA |
| Rock 15415 (Pure anorthosite) | 0.91 | 0.003 | RSA |
| Surveyor V (Tranquillitatis; soil) | 0.35 | | AST |
| Surveyor VI (Sinus Medii; soil) | 0.34 | 0.20 | AST |
| Surveyor VII (Rim of Tycho; soil) | 0.55 | 0.20 | AST |

[1]) Al/Si ratio varies considerably among Apollo 15 soils (range 0.24−0.43) mainly due to admixture from nearby mountains (rich in anorthosite). See[12].

landing sites of Apollo 11 and Apollo 15, and by the admixture of the third major component, KREEP, which is thought to be present in a layer beneath the anorthositic layer. We deal with KREEP in the next section.

## E. KREEP

During their excursion on the lunar surface the astronauts of Apollo 12 noticed lighter spots in the generally dark soil. This indicates the rather recent addition of foreign material. Foreign material was expected at this landing site, which was to some extent selected because one of the light-colored rays from the crater Copernicus crosses this site. All comparable large young craters show such rays. The most dominant ray crater is Tycho.

Chemical analyses at various laboratories showed that nearly all the soil samples brought back by Apollo 12 had a different chemical composition. Wänke et al.[6] assumed there to be only two components, present in varying ratios, and constructed a mixing diagram from the data (soil samples and breccias of Apollo 12) without making any assumptions about the composition of the components.

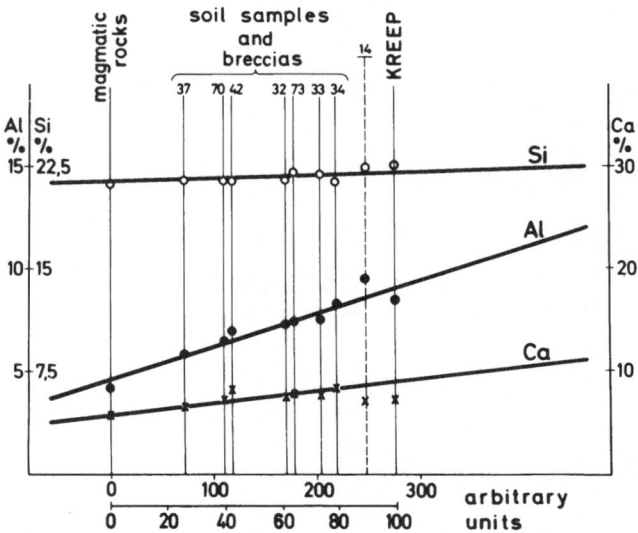

Fig. 7. Mixing diagram from the two-component model of Wänke et al.[6] for the Apollo 12 soils (12037, 12070, 12042, 12033) und breccias (12034, 12073). Note that the element lines and the compositions (position along the x-axis) were calculated via least-squares fit from the data for basaltic rocks, Apollo 12 soils and breccias only. The data for KREEP and the soil sample from the Fra Mauro formation (14) were not used for the computation of the element lines. The position of these samples along the x-axis was calculated via the previously determined element lines. Finally, the distance between the magmatic rocks and KREEP was divided into 100 units

As can clearly be seen from Figs. 7,8, and 9, the two-component assumption was correct, i.e. the contribution of other components is small. In addition it was realized that one of the two components was derived from the local mare basalts whereas the other was highly enriched in a number of elements. For example, potassium varied by a factor of almost 10. Various authors found millimeter-sized fragments of this foreign material which represents the second component in the Apollo 12 soil samples[28]. As the foreign component was rich in potassium, rare earth elements (REE) and phosphorus, the acronym KREEP was devised. Later many other elements were found to be enriched in KREEP, for example, U, Th, Hf, Zr, Nb, and Ta. All these elements have in common large crystal radii (large-ion lithophile elements, LIL elements). The rock type from which the KREEP fragments were derived of is norite.

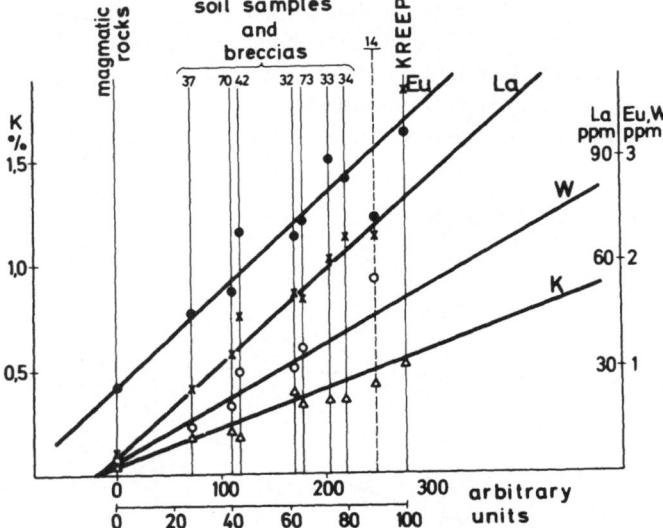

Fig. 8. See Fig. 7

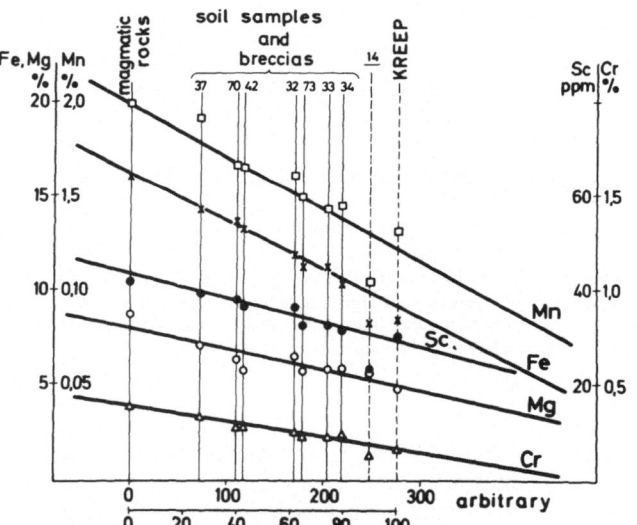

Fig. 9. See Fig. 7

As we will see in Chap. VI, we now have evidence that the KREEP (norite) component in the Apollo 12 soil represents ejecta from the crater Copernicus. We should note here that the composition of the Apollo 14 soil samples and KREEP is nearly identical. KREEP was later shown to be admixed also with the Apollo 15 and 16 soil samples. The Apollo 15 soil contains all three major components found so far: mare basalt, KREEP (norite), and anorthosite[12,16] (see Table 5).

Table 5. Concentrations of mare basalt, KREEP and anorthosite in Apollo 15 soils[12]. The highest anorthosite concentrations were found for the samples from locations close to the Apennine front

| Soil | Mare Basalt % | KREEP % | Anorthosite % | Station |
|------|---------------|---------|---------------|---------|
| 15601 | 86 | 9 | 5 | Hadley Rille |
| 15558 | 68 | 19 | 13 | Hadley Rille |
| 15501 | 71 | 15 | 14 | Scarplet Crater |
| 15021 | 62 | 20 | 18 | Lunar Module |
| 15471 | 71 | 11 | 18 | Dune Crater |
| 15301 | 62 | 15 | 23 | Spur Crater |
| 15265 | 41 | 28 | 31 | Front |
| 15271 | 48 | 22 | 30 | Front |
| 15101 | 49 | 17 | 34 | St. George Crater |

## III. The Europium Anomaly

One of the most striking findings during the analyses of the first lunar samples was the large negative europium anomaly found in all mare basalts [29–31] and later in the KREEP samples, too (Fig. 10).

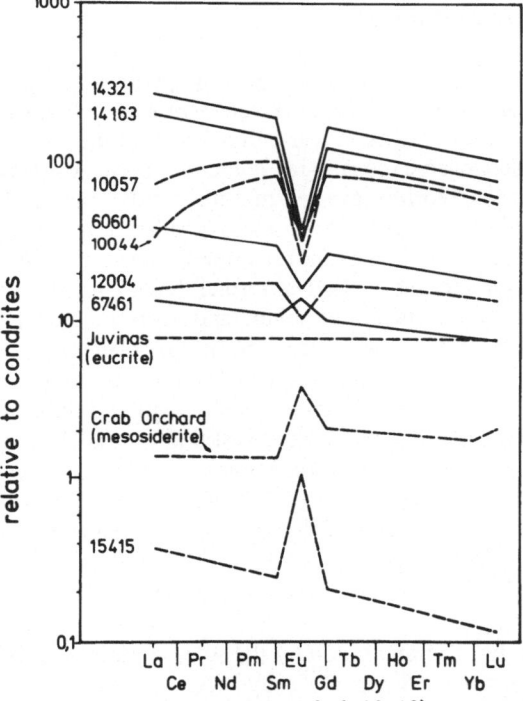

Fig. 10. Eu anomaly. All data from MPI, Mainz,[2, 6, 12, 13] except for rock 15415[8]

129

The absence of ferric iron in the lunar basalts indicates a low partial pressure of oxygen ($< 10^{-13}$ atmospheres). Under such highly reducing conditions europium occurs in the $2^+$ state rather than in the $3^+$ state. Thus, partial melting would concentrate the $Eu^{2+}$ in the feldspar (plagioclase) and the trivalent REE (rare earth elements) would tend to remain in the pyroxenes. Such a distribution was indeed found in mineral separates by Philpotts and Schnetzler[29]. Concentrations of the REE (with the exception of europium) were lower in all the separates than in the bulk sample. As reported by various authors, the REE are found in high concentrations in trace minerals.[32, 33] The phosphate minerals apatite and whitlockite seem to be of special importance for the mass balance of the REE. In fact the Eu anomaly has been attributed to the phosphates[34].

Positive Eu anomalies were found first in a few grains of Apollo 11 soil[35] and later on in various anorthosite samples where calcic feldspar (anorthite) is the dominant mineral.

As illustrated in Fig. 10, europium anomalies are found not only in lunar samples but also, for example, in the silicate phase of mesosiderites. This is strong evidence that these meteorites have undergone extensive magmatic fractionation. Of course, a low partial pressure of oxygen is required in order to keep europium in the $2^+$ state.

## IV. Solar Wind Implantation

Gerling and Levskii[36] were the first to report anomalously large amounts of noble gases in a stone meteorite. From the elemental abundance and the isotopic composition of these gases, it was quite clear that they are neither of radiogenic nor of spallogenic (cosmic-ray induced) origin. A survey of the noble gas content of stone meteorites revealed that many other meteorites contained an "excess" of noble gases; these were therefore called primordial gases[37-43].

Further studies showed that primordial noble gases are of two kinds: solar and planetary (Signer and Suess[39]). The former, being present in nearly solar proportions, i.e. consisting mainly of helium and neon and having $^{20}Ne/^{22}Ne$ ratios up to 14, were called solar noble gases. In the "planetary" component the heavier noble gases are enriched to the levels found in the terrestrial atmosphere. The origin of the planetary gases, which are present in nearly all chondrites and other types of stone meteorites, is still not quite clear, but they are commonly assumed to be derived in some way from the gas in the planetary nebula during the condensation process.[44] The solar noble gases were found to be located in the surface layers of the crystals of all major mineral components present in the meteorites in question[42, 45]. These findings led to the conclusions of Wänke[46, 47] that these noble gases originated from the solar wind.

One possible way of implantation by the solar wind into the solid grains which I discussed in my 1965 paper[47] is by irradiation of a layer of loose grains on the surface of a planetary body. Such a layer would be continuously mixed and turned over by the bombardment of meteorites down to the size of cosmic dust particles. I also stated that this body must not have any appreciable magnetic field nor an atmosphere and that the Moon would be a very suitable body for such a process to take place.

In line with these expectations, large amounts of noble gases (mainly helium) were found in the first lunar samples of Apollo 11 by the Lunar Sample Preliminary Examination Team (LSPET)[48]. The helium content of the lunar soils reached values of 0.2 $cm^3$ He/g, together with amounts of hydrogen up to about 1 $cm^3$ $H_2$/g. Beside the elemental and isotopic ratios, confirmation of the overwhelming contribution by the solar wind was provided by measurements on lunar fines of different grain sizes as well as by etching techniques, both of which showed the surface correlation of these gases[49-51].

Table 6 gives a few examples of noble gases in lunar samples derived from the solar wind. Concentrations in the microbreccias are comparable to those in the fines, indicating that the agglomeration process took place without much heating after the solar wind implantation.

Practically all the hydrogen in the lunar fines was implanted by the solar wind, as the indigenous water content of the lunar surface samples is extremely low. We can thus investigate solar hydrogen directly. Measurements indicated that the solar hydrogen was depleted in deuterium by a factor of at least 50[54]. Deuterium may even be completely absent in solar hydrogen since the amount present in the lunar fines could have been formed by spallation reactions.

The solar wind is again the major source of the carbon and nitrogen found in lunar fines. As noted first by Hintenberger et al.[49], nearly all of the nitrogen present in lunar fines is chemically bound (ammonium or nitride) and not in gaseous form ($N_2$)[55], as is most of the hydrogen, too. The nitrogen and hydrogen concentrations in the lunar fines and breccias are correlated with the grain surface area and are higher by at least a factor of 10 than in the lunar igneous rocks, which proves their solar wind origin.

Apart from the noble gases, hydrogen, carbon (see Eglinton[56]), and nitrogen, the solar wind contribution is negligible. Attempts to identify two favorable solar wind elements, sulfur and chlorine, were unsuccessful[2, 57].

The high content of argon-40 in lunar fines and breccias was a big surprise. The contribution of radiogenic argon-40 is likely to be small in most cases and that of spallogenic argon-40 is negligible. Although the argon-40, as well as argon-36 and the other noble gases, is correlated with the grain surface area, the argon-40 found cannot possibly have been supplied by the solar wind. The measured argon-40/argon-36 ratio is about 1; this is several orders of magnitude above the value predicated for the sun[58].

As suggested by Heymann et al.[59], the excess argon-40 must be of lunar origin. Argon-40 from the decay of potassium-40 in the interior of the moon must have escaped into the lunar atmosphere where it was ionized and driven back to the surface of the moon by interaction with the interplanetary electric and magnetic fields. Alternatively, the argon passing through the regolith on its way to the lunar surface was temporarily absorbed on the grain surfaces and then implanted by shock from impacts.

Table 6. Solar wind-derived rare gases in lunar and meteoritic samples

| Samples | $^4$He | $^{20}$Ne | $^{36}$Ar | $^4$He/$^3$He | $^{20}$Ne/$^{22}$Ne | $^4$He/$^{20}$Ne | Refs. |
|---|---|---|---|---|---|---|---|
| | in $10^{-6}$ cm$^3$ STP/g | | | | | | |
| *Lunar Soil* | | | | | | | |
| 10084 bulk | 190,000 | 2100 | 376 | 2550 | 12.4 | 91 | 49) |
| 10084 metal | 98,000 | 338 | 51 | 2530 | 12.4 | 290 | 49) |
| 10084 bulk (grain size 42 $\mu$) | 100,000 | 1200 | 168 | 2620 | 12.4 | 83 | 50) |
| 10084 ilmenite (grain size 41 $\mu$) | 389,000 | 1790 | 63.3 | 2570 | 12.9 | 217 | 50) |
| "0.16 $\mu$ removed | 295,000 | 941 | 24.6 | 2650 | 12.7 | 313 | 50) |
| "0.19 $\mu$ removed | 148,000 | 507 | 12.4 | 2440 | 12.5 | 292 | 50) |
| "0.35 $\mu$ removed | 37,800 | 130 | 3.6 | 2030 | 12.6 | 291 | 50) |
| 10087 bulk (grain size 25—35 $\mu$) | 85,000 | 1250 | 207 | 2740 | 12.8 | 68 | 52) |
| 12070 bulk (grain size 25—35 $\mu$) | 147,000 | 2430 | 427 | 2690 | 12.6 | 61 | 52) |
| *Lunar Breccias* | | | | | | | |
| 10021 bulk | 373,000 | 5620 | 745 | 2940 | 12.8 | 66 | 52) |
| 10021 bulk (grain size 25—35 $\mu$) | 168,000 | 2400 | 290 | 2930 | 12.5 | 70 | 52) |
| 10061 bulk | 255,000 | 5020 | 857 | 2980 | 12.5 | 51 | 52) |
| 10061 bulk (grain size 25—35 $\mu$) | 213,000 | 2840 | 466 | 2880 | 12.3 | 75 | 52) |
| *Chondrites* | | | | | | | |
| Fayetteville | 22,500 | 62.4 | 2.8 | 3400 | 12.3 | 360 | 53) |
| Pantar | 680 | 2.14 | 0.13 | 3200 | 12.5 | 320 | 39) |
| Pantar metal | 1,280 | 17.6 | — | 2200 | 13.3 | 800 | 45) |
| *Ca-rich achondrites* | | | | | | | |
| Kapoeta | 1,360 | 24 | 1.07 | 3600 | 14.0 | 57 | 37) |

As Heymann et al.[60, 61] have pointed out, after correction for $^{40}$Ar from radioactive decay of $^{40}$K, the $^{40}$Ar/$^{36}$Ar ratios may be correlated with the age of the surface rocks at the various landing sites (Table 7). This is plausible since the "age of the landing site" represents the earliest time at which $^{40}$Ar from the lunar atmosphere could have become implanted in the regolith. The escape of argon-40 has declined over about 4 billion years in parallel with the decay of potassium-40 and the amount of rock melting in the Moon. The regolith is well mixed at most places so the layers sampled by the astronauts always contain a certain proportion of grains that were on the surface early in the history of the regolith. Hence, these grains record the amount of argon-40 escaping from the lunar interior, i.e. the concentration of argon-40 in the lunar atmosphere.

Table 7. Argon $(^{40}Ar/^{36}Ar)_{tr}$ ratios taken from a summary of Heymann et al.[60] Rock datings from Wasserburg et al.[62]

| Soil Samples | Rock ages in b.y. | $(^{40}Ar/^{36}Ar)_{tr}$ range | average | Number of measurements |
|---|---|---|---|---|
| Apollo 14 | 3.90 | 0.8–1.9 | 1.3 | (3) |
| Apollo 11 | 3.65 | 0.7–1.4 | 1.1 | (9) |
| Luna 16 | 3.40 | – | 0.65 | (1) |
| Apollo 12 | 3.25 | 0.3–0.5 | 0.35 | (8) |

This explanation is probably confirmed by the results of Hintenberger et al.[52], who found that the ratio $(^{40}Ar/^{36}Ar)_{tr}$ is higher for breccias than for soil samples (Table 8). There is good reason to assume that most breccias were formed from loose soil when the regolith was relatively young[12]. Hence, the breccias should contain a higher percentage of "old" grains (that were on the surface early in the history of the regolith).

Table 8. Isotope ratios of trapped noble gases, cosmic ray exposure ages (in $10^6$ years) of soil samples and breccias (all data from Hintenberger et al.[52]. Isotope ratios were obtained by various methods (see original paper[52])

| Sample | Soils | | | Breccias | |
|---|---|---|---|---|---|
| | 10084 | 10087 | 12070 | 10021 | 10061 |
| $(^4He/^3He)_{tr}$ | $2620 \pm 10$ | $2780 \pm 40$ | $2740 \pm 10$ | $3070 \pm 20$ | $2930 \pm 20$ |
| $(^{20}Ne/^{21}Ne)_{tr}$ | $408 \pm 5$ | $409 \pm 6$ | $430 \pm 30$ | $408 \pm 4$ | $420 \pm 20$ |
| $(^{40}Ar/^{36}Ar)_{tr}$ | $0.95 \pm 0.02$ | $1.51 \pm 0.03$ | $0.44 \pm 0.01$ | $2.18 \pm 0.01$ | $1.81 \pm 0.02$ |
| $T(^{21}Ne)$ | $330 \pm 40$ | $430 \pm 40$ | $360 \pm 80$ | $600 \pm 70$ | $800 \pm 200$ |

## V. Meteoritic Component

Even before the first lunar landing, it was known that three types of meteoritic debris would be found on the Moon[63]:

1) micro-meteorites and small meteorites in the regolith
2) crater-forming bodies in ray material and other ejecta
3) planetesimals from the "early intensive bombardment" in ancient breccias and in regolith of the highlands (ancient component).[64]

The siderophile elements (Ir, Au, Re, Ni, etc.) seem to be the most reliable indicators of meteoritic material. These elements concentrate in the metal phase during the planetary melting and are therefore strongly depleted in the surface layers of differentiated planets (*e.g.* on Earth by a factor of $10^{-4}$). However, a number of volatile elements (Ag, Bi, Br, Cd, Ge, Pb, Se, Te, Zn) are so strongly depleted on the lunar surface that some of them can serve as subsidiary indicators of meteoritic matter.

Meteoritic matter on the Moon has been investigated mainly by Anders and coworkers[11, 65-68] and Wasson and coworkers[69-71]. Data on indicator elements were also obtained by various other authors[2, 6, 12, 13, 72-74].

Individual metal particles magnetically separated from lunar samples were studied intensively by Goldstein *et al.*[75-77], Wänke *et al.*[6, 78] and Wlotzka *et al.*[79, 80]. From their trace element composition as well as from their content of Ni and Co, it was clear that the majority of the particles are of meteoritic origin.

The gold content of type 1 carbonaceous chondrites is 0,15 ppm.[81] Anders *et al.*[68] assumed C 1 chondrites (the most primitive class of meteorites) to be the dominant contributing component and, after correction of the small amounts of siderophile elements found in the igneous lunar rocks, calculated from the data on Au given in Table 2 the meteoritic component as shown in Table 9. For all four mare sites the meteoritic components amount to 1.55–1.78%, evidencing a nearly constant admixture in the lunar regolith. Only slightly lower values are obtained when the highly volatile elements are used as indicators. Hence, the assumption that C 1 chondritic material is the meteoritic component is substantially correct, although some admixture of fractionated matter (less primitive meteorite classes) is seen, in the Apollo 11 and 12 soil samples, especially.

Table 9. Meteoritic component as calculated from the abundance of siderophile and volatile elements in bulk soils. In carbonaceous chondrite type 1 equivalent and best estimate of true C 1 component (from Anders *et al.*[68])

| Soils | Meteoritic component in percent | | Best estimate C 1 comp. |
|---|---|---|---|
| | C 1 equivalent | | |
| | siderophile | volatile | |
| Apollo 11 | 1.77 | 1.12 | 1.14 |
| Apollo 12 | 1.66 | 1.34 | 1.28 |
| Apollo 15 | 1.55 | 1.57 | 0.96 |
| Luna 16 | 1.78 | 2.23 | 1.6 |

The average thickness of the regolith is estimated 4.6 m[82], which gives an average rate of influx of meteoritic matter of $2.4 \times 10^{-9}$ g cm$^2$ yr$^{-1}$ for the three Apollo landing sites[68].

We do not yet know much about the crater-related component. Slight deviations from the C 1 abundance pattern are observed, especially in samples from the vicinity of Dune crater (Fra Mauro Formation). However, these could also be due to a mixture of C 1 with ancient component.

This ancient component has been observed in the highland soils, as predicted[67, 68]. Volatile elements do not give a clear picture, as the correction for the indigenous contribution to the ancient component is difficult to estimate. We shall therefore restrict the discussion to the siderophile elements, because the correction is likely to be quite small in this case. The old breccias of Apollo 14 had a meteoritic contribution with an Ir/Au ratio of about 1.5, as compared to about 3.5 in the mare soil samples. Moreover, the abundance pattern of the other indicator elements does not match that of any known meteorite class, as Ir, Re, and Ni are depleted relative to Au and As[13, 68, 80].

The composition of metal separates from various lunar and meteoritic samples is compiled in Table 10. The metal content of the lunar soil samples is about 0.5%, but somewhat rarer in the soil samples of Apollo 12 metal. Trace quantities of metal are found in the igneous rocks; the composition of these metals differs considerably from the meteoritic abundance pattern and is therefore easily distinguished. Of course, the metal from the lunar igneous rocks will also become admixed to soils. Hence, the metal separates of the soil samples are a mixture of meteoritic and indigenous metal from lunar rocks. As can be seen from Table 10, the latter contributes significantly only to metal from soil 15601. Measurements have been carried out on individual metal grains down to particles of less than $10^{-6}$ g weight via electron microprobe for Fe, Ni, and Co and via neutron activation for Fe, Co, Ni, Cu, Ga, As, W, Ir, and Au.[6]

The extremely high W content of the lunar metal was very puzzling at first, especially as it was clear that grains of meteoritic origin were the main contributors. Only metal from eucrites has a comparably high W content. Later it was shown by Wänke et al.[6] that W had been enriched in the early lunar differentiation processes. Actually W is one of the elements highly fractionated together with other large-ion lithophile elements (LIL elements). The highest concentration of LIL elements is found in KREEP (norite).

During impacts, the meteoritic metal equilibrates with the surrounding silicate at elevated temperatures and W, having considerable siderophile tendency, enters the metal phase.

For the siderophile elements, the metal particles contribute between 50 and 90% to the bulk composition of the soil samples. As carbonaceous chondrites do not contain metal, reduction and equilibration is required to explain the high contribution of the metal particles.

Table 10. Composition of lunar and meteoritic metal particles. Bulk composition of C 1 is given for comparison. In all cases – except C 1 – difference to 100% equal Fe

| % | Soil | | | | | Rock | Pultusk | Ramsdorf | Juvinas | C 1 |
|---|---|---|---|---|---|---|---|---|---|---|
| | 10084 | 12001 | 14163 | 15601 | 60601 | 12053 | H-chondr. | L-chondr. | eucrite | bulk |
| Ni | 4.7 | 6.4 | 5.7 | 4.9 | 6.1 | 1.9 | 9.1 | 13.9 | 2.9 | 0.98 |
| Co | 0.52 | 0.65 | 0.55 | 1.04 | 0.37 | 1.36 | 0.46 | 0.63 | 0.62 | 0.050 |
| ppm | | | | | | | | | | |
| Cu | 340 | 290 | 220 | 860 | 260 | 675 | 430 | 380 | 150 | 130 |
| Ga | 11 | 14 | 14 | 9 | – | 9 | 18 | 32 | 9 | 12 |
| Ge | – | – | 130 | – | – | – | – | – | – | 31.2 |
| As | – | – | 11 | – | 18 | – | – | – | – | 1.8 |
| Pd | – | – | 2.8 | – | – | – | – | – | – | 0.5 |
| W | 24 | 110 | 223 | 25 | 48 | 33 | 1.1 | 0.7 | 21 | 0.1 |
| Ir | 2.6 | 5.5 | 2.0 | – | 1.6 | 0.1 | 4.3 | 4.9 | 0.9 | 0.51 |
| Au | 0.83 | 1.04 | 1.1 | 0.35 | 1.26 | 0.5 | 1.18 | 1.65 | 0.76 | 0.15 |
| Refs. | 78) | 6) | 79) | 79) | 80) | 6) | 78) | 78) | 78) | 17, 81) |

## VI. Formation of the Lunar Landscape

We can restrict this discussion to the front side of the Moon for two reasons. First, the back side of the Moon does not show any specific features. All the large maria are located on the front — a fact which is still not quite understood — while the back is similar in appearance to the highland areas of the front side. As we have seen from Table 4, this similarity holds true for the chemical composition, too. (The first three lines in Table 4 refer to locations on the lunar back side). Second, our knowledge of the lunar back side is based on photographs and a few isolated observations by the Apollo Command Modules.

From the front side we now have exact solidification ages for the rocks of the major maria. The numbers in Table 11 refer to the most recent lava flows detected and represent the time of solidification of the outermost layer.

Table 11. Solidification ages of the lunar maria as derived from Rb/Sr age determinations (isochron ages) by Wasserburg and coworkers

| Maria | Mission | Age | Number of rocks analyzed | Refs. |
|---|---|---|---|---|
| Mare Tranquillitatis | Apollo 11 | 3.65 ± 0.06 b.y. | 7 | 83) |
| Oceanus Procellarum | Apollo 12 | 3.26 ± 0.10 b.y. | 8 | 84) |
| Mare Imbrium | Apollo 15 | 3.36 ± 0.07 b.y. | 6 | 62) |
| Mare Fecunditatis | Luna 16 | 3.42 ± 0.18 b.y. | 1 | 62) |

There seems to be fairly general agreement among scientists that at least Mare Imbrium and the other circular maria were originally large basins formed by impact.[21] The lava filling extended over several hundred million years and was not restricted to the diameter of the origin crater, the surrounding areas being covered, too.

According to the model of Wood[85], the early differentiation of the upper 100 to 200 km of the Moon resulted in a light anorthositic crust while the mafic magnesium-rich minerals would have sunk towards the bottom of the molten layer. Between the two layers, or just below the anorthositic crust, a layer highly enriched in various trace elements (KREEP) was formed. These rocks are called norites. Intensive bombardment probably hindered complete segregation of anorthosite and norite, or resulted in a considerable mixing of the two layers. Perhaps the norite was still hot and partially molten at this stage; thus, the lava flows, which are seen in the highlands in connection with major craters might be noritic.

After solidification of the lunar crust, the Imbrium collision was the most spectacular event. About $3.9 \times 10^9$ years ago an object with an estimated diameter of 200 km but probably having a rather low velocity of 3 to 5 km/sec hit the surface of the Moon, which must already have been similar in appearance to the highland areas as we see them today.[86] The impact excavated a basin about 700 km in diameter and 100 km deep. The anorthosite layer, which may have been rather thin at this locality, was blown away completely and norite material from beneath was thrown out and deposited over wide areas on the

137

front side of the Moon. A triple ring wall structure was formed like that known from Mare Orientale on the back side of the Moon.

The model described here was predicted by Baldwin[87] and especially by Urey[86] on the basis of the structural features of the surrounding areas, as first pointed out by Gilbert in 1893[88]. One example is the Fra Mauro Formation, first interpreted by Eggleton[89,90] as part of a thick blanket of ejecta from the Imbrium basin. This interpretation has been confirmed by the investigations on the Apollo 14 samples[91,92].

Later on, upwelling lava from the deep lunar interior filled the Imbrium basin. The inner walls of the ring mountains collapsed and only the outer wall remained. Montes Apenninus and Carpatus are part of the outer wall. The lava flowed over the whole area, forming the nearly circular Mare Imbrium with a diameter of about 1150 km. The surrounding area probably subsided and was later covered with lava which found its way through the outer wall.

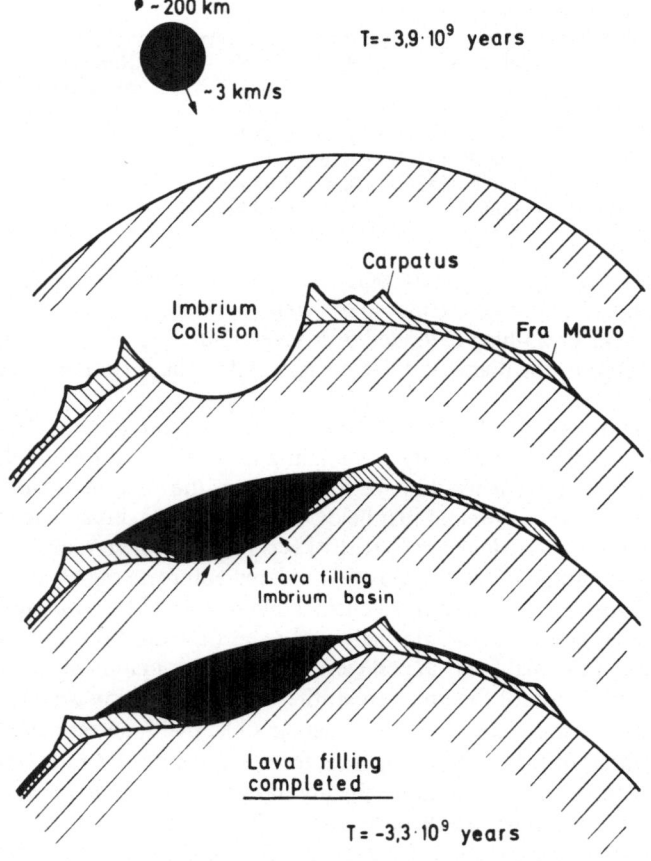

Fig. 11. Simplified sketch of the Imbrium collision and lava filling of Mare Imbrium and Oceanus Procellarum

The lava filling was completed about $3.3 \times 10^9$ years ago. The rocks from Oceanus Procellarum (Apollo 12) and from the edge of Mare Imbrium (Apollo 15), about 1000 km apart, both solidified at this time. The elevated area of the Fra Mauro Formation was not covered with lava.

Here, we still find noritic material on the surface. Age determinations on selected Apollo 14 samples showed that the material of the Fra Mauro area underwent thermal metamorphism about $3.9 \times 10^9$ years ago[93]. The Imbrium collision is thought to have provided the required energy and is therefore dated accordingly by various authors[93,94,95].

The lava layer of Oceanus Procellarum is estimated to have a maximum thickness of a few hundred meters, covering the noritic ejecta from the Imbrium impact. At least part of the foreign (KREEP) component from the Apollo 12 soils described in Chap. II. D was severely heated about 850 million years ago[96,97]. Most authors agree that this material consists of ejecta from Crater Copernicus[63].

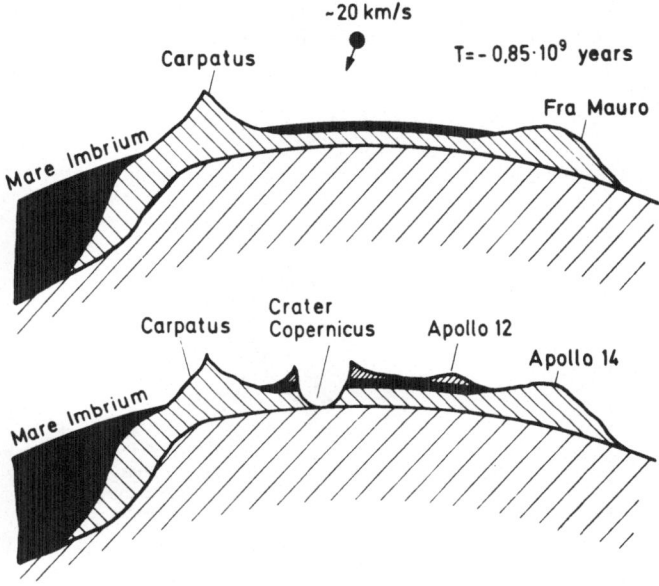

Fig. 12. Copernicus impact throwing material deposited there by the Imbrium collision to the landing site of Apollo 12

If this conclusion is correct (for the contrary view, see Wasson and Baedecker[98]), Copernicus is 850 million years old, which is not in contradiction with estimates from crater counting[99]. The impact penetrated the thin basalt layer and deeply excavated the norite material deposited there by the Imbrium event; alternatively, the surface layer may have consisted of noritic material at this locality.

## VII. Element Correlations

Many element correlations were discovered in the lunar samples, the first among a group of elements found in KREEP.

In principle, correlated elements must show similar geochemical behavior. Two somewhat different processes are involved:

a) Elements similar in ionic radius and valency tend to enter the same major rock-forming minerals.

b) Elements which do not readily enter the major phases concentrate in the residual liquids and are therefore found in different trace minerals.

Examples of both types are shown in Figs. 13, 14, and 15. Clearly the correlation of MnO with FeO, found in the lunar samples by Laul *et al.*[100], is due to the fact that $Mn^{++}$ ($R = 0.80$ A) can easily replace $Fe^{++}$ ($R = 0.74$ A) in the two most abundant Mg, Fe silicates, pyroxene and olivine. The correlation of the LIL elements (large-ion lithophile elements) first observed in KREEP is of the second type.

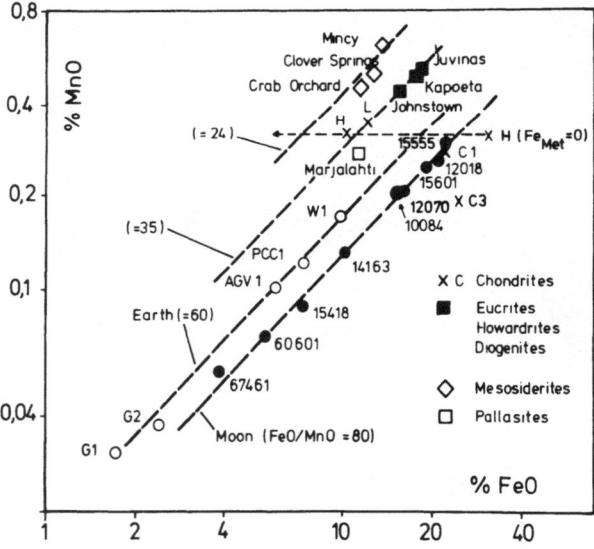

Fig. 13. MnO/FeO correlation first noted by Laul *et al.*[100] for lunar samples. A similar correlation seems to hold for terrestrial and meteoritic samples, too. By oxidation of metallic iron the data point of a material with the composition an ordinary H-group chondrite (marked H) will move along the horizontal dashed line. The location of the point of maximum oxidation ($Fe_{metal} = 0$) is also indicated

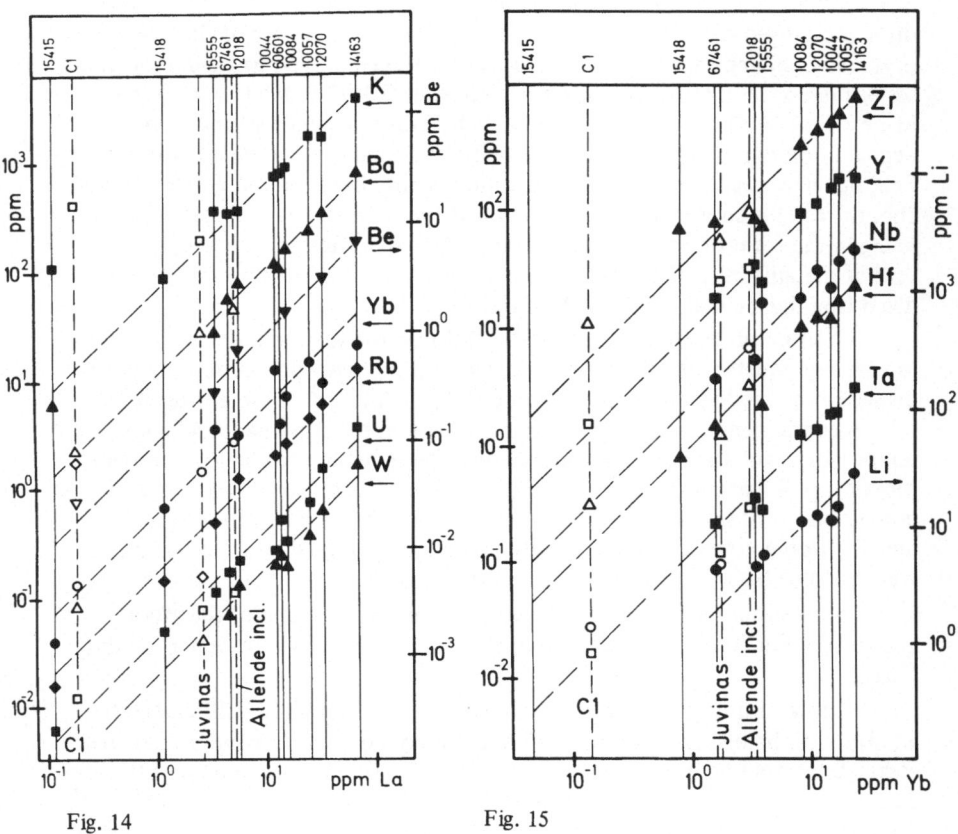

Fig. 14                                   Fig. 15

Fig. 14. Correlation of large-ion lithophile elements (LIL) following lanthanum in lunar samples. Note that Be ($R = 0.35$ A) fractionates together with the LIL elements. From Wänke et al.[13]

Fig. 15. Correlation of LIL elements following ytterbium in lunar samples. Note that Li ($R = 0.68$ A) seems to fractionate together with the LIL elements. From Wänke et al.[13]

Correlations between pairs or groups of elements are of extraordinary significance, as from a limited number of samples collected at the surface elemental ratios can be determined for a whole planet, or at least for that fraction of the planet which underwent magmatic differentiation.

In all lunar samples of MnO and FeO are correlated exactly (Fig. 13), the FeO/MnO ratio being 80. A similar trend is seen to exist for the Earth, and various groups of meteorites. The FeO/MnO ratio is lowest for the silicate phase of mesosiderites (FeO/MnO = 24), somewhat higher (35) for eucrites, howardites and diogenites (all achondrites) and for the silicate phase of the low-iron (hypersthene) and high-iron (bronzite) chondrites (the two most common classes of meteorites). The carbonaceous chondrites C 1, C 2, and C 3 have the highest FeO/MnO ratio. Clearly the FeO/MnO ratio depends strongly on the degree of oxidation. Carbonaceous chondrites contain only oxidized iron and FeS, but no metallic iron; they are the most highly oxidized meteorites. In chondrites of groups H and L, less than half the total iron is oxidized, the rest being in the form of FeS and metallic iron. Mesosiderites have the highest content of metallic iron. The highly reduced enstatite chondrites contain no oxidized iron, hence their FeO/MnO ratio is zero.

The situation is complicated by the fact that the total amounts of Fe and Mn are not constant for the various planets and the individual groups of meteorites. In the models proposed by Wood[101] and Anders[102] meteorites consist of two components: a high-temperature, volatile-free component and a low-temperature, volatile-rich component. Manganese belongs to the low-temperature group. Recent studies[103, 104] indicate that at least one additional component has to be introduced into the model, namely metallic iron, because metallic iron condenses independently, as do some other siderophile elements. We will come back to these questions in the next chapter.

Among other facts, the difference in the FeO/MnO ratio for the Moon and for the Earth rules out the possibility that the Moon was once part of the Earth.[100]

We now turn to the correlations shown in Figs. 14 and 15. LIL elements are found to be highly enriched in all KREEP-containing soil samples, but the correlations hold for the igneous rocks as well.[13, 14] Only the pure anorthosites deviate significantly. As already seen in Fig. 10, lanthanum and the heavier REE (rare earth elements) fractionate somewhat differently. The elements plotted in Figs. 14 and 15 fall roughly into two groups, one behaving like lanthanum and the other like ytterbium. This distinction means that the data points for individual samples lie closer to the correlation lines. Nevertheless, if all elements shown in Fig. 15 were plotted against lanthanum the entire range would show good correlation, as can be imagined from the Yb-La correlation in Fig. 14.

Refractory elements, i.e. REE, Be, Y, Zr, Hf, Nb, and Ta, are present on the Moon in their solar (C 1) abundance ratios. Only W is considerably depleted relative to the other refractories, indicating the presence on the Moon of metallic iron, which is responsible for the depletion of this rather siderophile element.

The alkali metals Li, K, Rb, and Cs (Na does not fractionate together with the LIL elements) are highly depleted on the Moon relative to the refractory elements. Lithium is less depleted than K, Rb, and Cs.

The values derived from these correlations for the elemental abundance ratios should be representative for the whole Moon. On this basis it is not possible to distinguish between depletion of the more volatile elements (K, Rb, and Cs) or enrichment of the refractories. As we will see later, both processes have occurred.

Fig. 14 includes terrestrial and meteoritic samples which show that the good correlation of La and K is not restricted to the Moon. It is interesting to note that the correlation line for the Moon falls close to that of the basaltic achondrites and that furthermore the silicate phases of mesosiderites and the hypersthene chondrite (diogenite) Johnstown fall on the line of the basaltic achondrites.

## VIII. Condensation Sequence of the Planetary Nebula

We have already discussed how the elements condense in accordance with their condensation temperatures and their chemical affinities in at least three distinguishable groups. Condensation sequences have been calculated by Urey[105, 106], Lord[107], Larimer[108] and others. According to Larimer[108] and Anders[109, 110] a cooling gas of cosmic composition should condense in the following order: below 2000 °K the highly refractory compounds of Ca, Al, Mg, Ti, and Si appear, followed by magnesium silicates and iron at 1350 to 1200 °K, and alkali silicates at 1100 to 1000 °K. Only H, C, N, O, S, and some volatile trace elements still remain in the gas phase. At 680 °K $H_2S$ begins to react with metallic Fe grains forming FeS. The elements Pb, Bi, Tl, and In condense between 600 and 400 °K. At 400 °K $H_2O$ reacts with the remaining metallic iron to give $Fe_3O_4$. At still lower temperatures water can be bound as hydrated silicates.

From new data Grossman[103] has recently shown that iron-nickel alloys have higher condensation temperatures than the Mg silicates at all total pressures above $7.1 \times 10^{-5}$ atmospheres. The estimated pressure in the primitive solar nebula in the Earth's orbit is around $10^{-3}$ atmospheres[111]. The temperature difference between the condensation points of metallic iron and Mg silicates increases with pressure allowing the possibility of greater fractionation of metal from silicate towards the center of the solar nebula where the pressure and temperature were highest. Hence, iron probably condensed to a large degree independently of the Mg silicates[103].

Grossman[112], thinks that the Ca, Al rich inclusions found in C 3 chondrites represent these high-temperature condensates (HTC). In Ca, Al-rich inclusions of the Allende meteorite he found Sc, La, Sm, Eu, Yb, and Ir together with Ca and Al to be enriched by a factor of about 20 relative to C 1 chondrites.

The work of Grossman[112] was extended in this laboratory[113]. Together with earlier results on individual elements, we can now present the following list of elements, all enriched in the Allende inclusions by a factor of about 20 relative to C 1 chondrites: Al, Ca, Ti, Sc, Sr, Y, Zr, Nb, Mo, Ru, Ba, La, Ce, Pr, Sm, Eu, Gd, Dy, Ho, Er, Yb, Lu, Hf, Ta, W, Re, Os, Ir, Pt, U, and probably Th. The inclusions also contain appropriate amounts of Mg, Si, and O, but most of the other elements are highly depleted.

The trace elements do not condense strictly according to their condensation temperatures of the pure elements or their components. For example, in spite of their considerably lower condensation temperatures the REE elements are found in the Allende inclusions in their true cosmic ratio relative to one another and relative to the very refractory elements. Solid solution effects can be expected to allow the condensation of the REE elements long before saturation is reached. The second major phase to condense, perovskite ($CaTiO_3$), could provide lattice sites for the removal of the REE and other trace elements from the gas phase[112]. Similarly, Au, Ge and other siderophile elements with lower condensation temperatures may be expected to condense together with the Fe-Ni alloy[110].

At any stage of condensation the fraction still in the gas phase can be dispersed to other regions of the planetary system. Hence, the accumulating planetary objects may contain different components in varying amounts. As can be seen from Fig. 16, among all objects of which we have samples, the Moon contains the highest proportion of HTC elements, closely followed by several groups of achondrites.

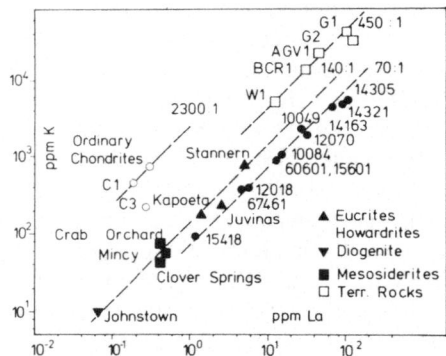

Fig. 16. K *vs.* La in lunar, terrestrial, and meteoritic samples. From Wänke *et al.*[13]

## IX. Composition and Structure of the Moon

### A. Chemical Composition

From the correlated elements we can construct a model of the chemical composition of the Moon. Whether this model is valid for the entire Moon or only for the upper 500 km, depends on the amount of melting that has occurred in the Moon. Clearly, if below a certain depth there has never been any matter exchange with the upper layers, it is impossible to deduce anything about the composition of this deep-seated matter from studies of the surface rocks.

The high value for the heat flux through the surface of the Moon measured by Langseth *et al.*[114] requires high concentrations of radioactive elements and therefore clearly favors extensive melting of large parts of the Moon. It should be borne in mind that the upper 500 km of the Moon comprises about two thirds of its total mass.

In the preceding chapter we presented the proof that on the Moon the refractory elements are highly enriched relative to the more volatile ones.

A lower K/U ratio for the Earth and the basaltic achondrites as compared to the chondritic (solar) ratio was emphasized more than ten years ago by Gast[115, 116] and Wänke[117]. Since the first lunar samples became available it has been suggested, especially by Gast[30, 118], that the Moon has a higher concentration of the refractory elements Ca, Al, REE, etc.

Anderson[119] has proposed that the Moon has a composition similar to that of the Allende inclusions, as this would explain the low values observed for electrical conductivity[120] of the lunar interior in spite of the estimated high temperatures. Other authors have proposed for various reasons that the Moon may be eucritic. As pointed out by Laul et al.[100], a true eucritic or howarditic Moon is excluded by the difference in the FeO/MnO ratios.

The existence of the high-temperature condensates and their composition has in the meantime been well established by work on the Allende inclusions. We have convincing evidence that this high-temperature component (HTC) is responsible for the enrichment of the refractory elements, both for the Moon and the basaltic achondrites, in spite of the fact that the absolute percentage of the HTC is somewhat different.

As the HTC does not contain low-temperature elements like potassium etc., we assume that the Moon and the other objects whose K/La ratio differs from the chondritic value consist of a mixture of HTC and a component containing low-temperature elements.

Wänke et al.[13] have proposed a model in which the Moon is composed of a mixture of HTC and a material equivalent in elemental composition to bronzite (H = high-iron group) chondrites, but with a different degree of oxidation. Evidence in favor of ordinary chondrites as the component containing the low-temperature elements is presented in Fig. 17. Normalized to potassium, the ratios of the alkali metal elements in the lunar soil samples 12070 are plotted relative to the C 1 chondrites as well as relative to the H-group chondrite Forest City. We have previously seen that K, Rb, and Cs are correlated with one an-

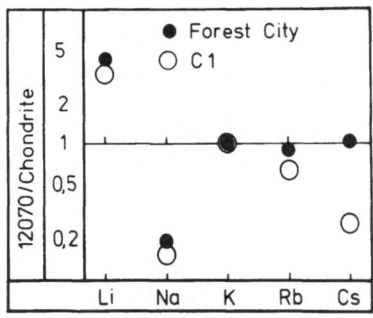

Fig. 17. Concentrations of alkali elements in soil sample 12070 relative to carbonaceous chondrites type 1, and relative to Forest City, an ordinary chondrite (bronzite chondrite). Data from Wasserburg's group[121, 122] and Mason[17]

H. Wänke

other, but not Li and Na. However, the concentrations of Rb and Cs in Forest City are in fact more similar to those of C 1 chondrites than to those of ordinary chondrites.

With the above assumption, it is easy to calculate the absolute elemental abundances of the Moon from the observed K/La ratio in the lunar samples and the concentrations of K and La both in the HTC (as analyzed in the Allende inclusions) and in normal chondrites.

The mixing ratio of HTC and chondritic component can be calculated from the following relation:

$A$ = chondritic fraction
$(1-A)$ = HTC fraction
indices:

m = moon
a = chondrites
b = HTC

$$\frac{[K]_m}{[La]_m} = R = \frac{A[K]_a + (1-A)[K]_b}{A[La]_a + (1-A)[La]_b}$$

With a value for the concentration ratio $R = [K]_m/[La]_m$ of the lunar samples of 70, $[K]_b = 0$, $[K]_a = 800$ ppm, $[La]_a = 0,35$ ppm and $[La]_b =$

Table 12. Elemental composition of the Moon calculated according the model of Wänke et al.[13]

|  | HTC | HTC x 0.69 | Chondritic Comp. | Chond. Comp. x 0.31 | Moon calcul. |
|---|---|---|---|---|---|
| **%** | | | | | |
| Mg | 6.5 | 4.5 | 14.2 | 4.4 | 8.9 |
| Al | 17 | 11.7 | 1.0 | 0.3 | 12.0 |
| Si | 14 | 9.7 | 17.1 | 5.3 | 15.0 |
| Ca | 18 | 12.4 | 1.2 | 0.4 | 12.8 |
| Ti | 0.91 | 0.63 | 0.21 | 0.07 | 0.70 |
| Fe | – | – | 27.6 | 8.6 | 8.6 |
| **ppm** | | | | | |
| Na | – | – | 5700 | 1770 | 1770 |
| K | – | – | 800 | 250 | 250 |
| Rb | – | – | 2.8 | 0.87 | 0.87 |
| Cs | – | – | 0.107 | 0.033 | 0.033 |
| Mn | – | – | 2260 | 700 | 700 |
| Sr | 130 | 90 | 10 | 3.1 | 93 |
| Ba | 47 | 32 | 3.5 | 1.1 | 33 |
| Sc | 114 | 79 | 7.5 | 2.3 | 81 |
| Y | 33 | 23 | 2.2 | 0.7 | 24 |
| La | 4.9 | 3.4 | 0.35 | 0.1 | 3.5 |
| Eu | 1.14 | 0.79 | 0.078 | 0.02 | 0.81 |
| Zr | 93 | 64 | 10 | 3.1 | 67 |
| Hf | 3.3 | 2.3 | 0.2 | 0.06 | 2.4 |
| Nb | 6.6 | 4.6 | 0.5 | 0.16 | 4.8 |
| Ta | 0.3 | 0.2 | 0.025 | 0.008 | 0.2 |
| Mo | 6 | 4 | 1.7 | 0.5 | 4.5 |
| W | 1.84 | 1.30 | 0.15 | 0.05 | 1.35 |
| U | 0.12 | 0.083 | 0.011 | 0.003 | 0.086 |

4.9 ppm, we get 69% HTC, the remainder having chondritic composition.
For the percentage of the HTC in the achondrites (K/La = 140) of Fig. 16,
a value of 48% is obtained; for the Earth with K/La = 450, it is 22%. From
this we calculated the elemental abundances for the Moon. They are summarized in Table 12.

The content of metallic iron can be estimated from the FeO/MnO ratio
which is 80 in all lunar samples (Fig. 13). If we assume 904 ppm MnO for the
whole Moon (see Table 12), we get 7.2% FeO. The content of FeS is assumed
to be 0.31 x the concentration found in normal chondrites = 1.7% FeS. Sub-
tracting the iron in FeO and FeS from the total iron content, we finally get
for the content of metallic iron in the Moon $Fe_{met}$ = 1.9%.

For a number of elements the concentrations found in all lunar samples
varied by a factor of less than 3. In Table 13 we have compared these concentrations with the calculated values for the whole Moon.

Table 13. Comparison of the calculated element
concentrations with those found in lunar samples.
Only those elements are listed which show variation
of less than a factor of 4

|    | Observed range | Calcul. Table 12 |
|----|----------------|------------------|
| Ca | 5.1–14.1       | 12.8             |
| Na | 1430–5000      | 1770             |
| Eu | 0.73–2.7       | 0.81             |
| Sr | 74–244         | 93               |
| Sc | 40–92[1])      | 81               |

[1]) Mare basalts only.

The estimated concentrations for the heavy alkali metal elements Rb and
Cs are somewhat uncertain due to their variable content in normal chondrites.
Nevertheless, the ratio Rb/Sr = 0.009 derived in this calculation is in accord
with the value of ~ 0.005 postulated for the whole Moon by Papanastassiou
and Wasserburg[123].

Assuming that the Th content is 3.6 times that of uranium, we can calculate
heat production in the lunar interior due to the decay of the radioactive elements K, Th, and U. The value obtained for the heat flux agrees exactly with
the $0.79 \times 10^{-6}$ cal/cm$^2$ sec measured on the Moon as reported by Langseth
et al.[114].

## B. Internal Structure

Seismic experiments[124] indicate that today the Moon must be solid down to
a depth of 800 km, but they also suggest a small molten core. The positive
gravity anomalies found for the circular maria also require a thick, solid lunar
crust or mantle. If the whole Moon had a homogeneous distribution of radio-

elements with concentrations as given in Table 12, the lunar interior would still be completely molten. These constraints can only be removed by assuming that the radioelements are concentrated in shallow layers, thus allowing, the heat produced to be dissipated to the surface.

Wood[85] has pointed out that, due to an "important accident" of geochemistry, the heat-generating elements K, Th, and U tend to concentrate in the first eutectic liquid that appears when a rock begins to melt. If as is generally the case, the melt is less dense than the residual solids, and tends to be driven towards the surface of the planet, it carries part of the heat-generating potential with it. This can profoundly affect the thermal history of a planet[125].

Had the Moon been originally homogeneous, extensive partial melting would have occurred soon after accretion. The high concentration of radioelements (Table 12) would result in a temperature increase of about 500 °C within $3 \times 10^8$ years[12]. Furthermore, we can assume that a substantial fraction of the accretional energy has been retained as heat.

The melt, containing large fractions of radioelements and other LIL elements from the Moon's interior, will move towards the surface, forming a liquid layer of perhaps 100 km thickness. On cooling, this liquid will differentiate as outlined by Wood[85]. A solid crust of anorthosite will form on top and will increase in thickness, while the denser mafic cumulates will concentrate at the bottom. In the liquid layer between the anorthosite and the mafic cumulates the radioelements, together with all large-ion lithophile elements, will become further enriched. From this layer the KREEP (norite) material is derived.

The evidence from seismic experiments is fully in agreement with the model outlined above. In the Fra Mauro region of Oceanus Procellarum the crust is about 65 km thick and overlies a solid mantle[126]. The crust, in turn, is divided into two parts covered with a thin regolith. For comparison with the returned lunar samples, the upper layer has been inferred to be mare basalt and the lower one anorthositic gabbros.

The very high apparent seismic wave velocities of about 9 km/sec obtained for the layer below the crust imply that the lunar mantle is differentiated and its composition varies with depth.

A crucial test of the validity of lunar models is their agreement with the mean density and the coefficient of the moment of inertia of the Moon. The best values are:[127] $\bar{\rho} = 3.361 \pm 0.002$ g/cm$^3$ and $K \equiv C/MR^2 = 0.402 \pm 0.002$, which is almost identical with the value for the exact sphere $2/5 = 0.4000$. Ringwood and Essene[128] have argued that the lunar moment of inertia and mean density exclude a composition in the lunar interior similar to that of the mare basalts, as such basalts would transform to eclogite, resulting in excessively high density. In contrast, Gast and Giuli[129] have presented evidence that high-density eclogite ($\rho \sim 3.5$ g/cm$^3$) in the upper mantle is necessary in order to counterbalance the low-density crust ($\rho \sim 2.9$ g/cm$^3$) and to keep the coefficient of the moment of inertia at the observed value.

## X. The Origin of the Moon

Three major mechanisms are postulated to explain the origin of the Moon, and have long been the subject of discussion:

a) fission of the Earth
b) independent formation of Earth and Moon as twin planets
c) independent formation of the Moon somewhere in the solar system followed by capture by the Earth.

Formation by precipitation from the hot gaseous silicate atmosphere escaped from the Earth, as proposed by Ringwood [130], may be considered a variant of the common fission hypothesis (Model a). Fission can be excluded because of the arguments presented in Chap. VII. There are a number of severe constraints that are difficult to overcome in both the precipitation hypothesis and the twin-planet model.

Recently, the capture hypothesis has gained fresh support despite the generally known fact that the capture probability of the Moon is very low. One can, of course, always argue that, as there is only one Moon, statistical laws do not apply.

The most recent estimates of the densities of the inner planets give the following values [131]: Mercury 5.42, Venus 5.25, Earth 5.51, and Mars 3.96. Corrected to zero pressure, the numbers change to: Mercury 5.3, Venus 4,40, Earth 4.45, and Mars 3.85. As pointed out by Lewis [104], this density sequence could be attributed to accretion of these four planets at different temperatures. In accordance with the discussion presented in Chap. VIII, Mercury would consist mainly of the high-temperature component and metal, whereas components like $MgSiO_3$ condensing below the Fe-Ni alloy would be rare. Venus would have its cosmic share of all elements condensing above the temperature of formation of FeS (about 600 °K). The fact that the Earth has a higher density than Venus can be attributed to the reaction of metallic iron with $H_2S$ to form FeS. Finally, Mars would have accreted at a temperature at which nearly all the iron is oxidized, resulting in a further drop in density.

In line with this model, Cameron [132] has placed the origin of the Moon inside the orbit of Mercury. At this location the HTC dominates. Cameron further suggests that the Moon was thrown into an elliptic orbit by a close encounter with Mercury and later caught by the Earth. The energy required for the orbital change of the Moon was supplied by Mercury and as a consequence this planet was brought into its eccentric orbit. Cameron takes the high eccentricity of the present orbit of Mercury as further cue for his hypothesis, which, though it may sound highly speculative, fits the observations beautifully. One might further speculate that the Moon acquired its HTC at the proper place, inside the orbit of Mercury, but that the chondritic fraction was added only after the change of orbit when the Moon swept up material condensing around one astronomical unit.

H. Wänke

The picture presented here of the composition and origin of the Moon is liable to change as science advances. In any case, three years of research on the Moon and on the lunar samples have brought about, a drastic change in our understanding of the composition and development of the solar system and the nearest surroundings of its most beautiful and exciting planet, Earth. This gain in fundamental knowledge about the objects external to it help in a most remarkable way to unravel the secrets of the Earth.

*Acknowledgements.* I wish to thank Dr. H. Palme for many valuable suggestions and Miss H. Prager for her dedicated help in preparing this manuscript.

## XI. References

1) Turkevich, A. L., Franzgrote, E. J., Patterson, J. H.: Science *165*, 277 (1969).
2) Wänke, H., Rieder, R., Baddenhausen, H., Spettel, B., Teschke, F., Quijano-Rico, M., Balacescu, A.: Proc. Apollo 11 Lunar Sci. Conf., Suppl. 1, Geochim. Cosmochim. Acta, Vol. *2*, p. 1719 Pergamon Press (1970).
3) Franzgrote, E. J., Patterson, J. H., Turkevich, A. L.: Science *167*, 376 (1970).
4) Patterson, J. H., Turkevich, A. L., Franzgrote, E. J., Economou, T. E., Sowinski, K. P.: Science *168*, 825 (1970).
5) Öpik, E. J.: Ann. Rev. Astronomy Astrophys. *7*, 473 (1969).
6) Wänke, H., Wlotzka, F., Baddenhausen, H., Balacescu, A., Spettel, B., Teschke, F., Jagoutz, E., Kruse, H., Quijano-Rice, M., Rieder, R.: Proc. Second Lunar Sci. Conf., Suppl. 2, Geochim. Cosmochim. Acta, Vol. *2*, p. 1187. M. I. T. Press (1971).
7) LSPET (Lunar Sample Preliminary Examination Team): Science *175*, 363 (1972).
8) Hubbard, N. J., Gast, P. W., Meyer, C., Nyquist, L. E., Shih, C.: Earth Planet. Sci. Lett. *13*, 71 (1971).
9) Laul, J. C., Wakita, H., Schmitt, R. A.: Bulk and REE abundances in anorthosites and noritic fragments. In: The Apollo 15 Lunar Samples (eds. J. W. Chamberlain and C. Watkins) p. 221. Houston: Lunar Science Institute (1972).
10) Ganapathy, R., Keays, R. R., Laul, J. C., Anders, E.: Proc. Apollo 11 Lunar Sci. Conf., Suppl. 1, Geochim. Cosmochim. Acta, Vol. *2*, p. 1117 Pergamon Press (1970)
11) Morgan, J. W., Krähenbühl, U., Ganapathy, R., Anders, E.: Proc. Third Lunar Sci. Conf., Suppl. 3, Geochim. Cosmochim. Acta, Vol. *2*, p. 1361. M. I. T. Press (1972).
12) Wänke, H., Baddenhausen, H., Balacescu. A., Teschke, F., Spettel, B., Dreibus, G., Palme, H., Quijano-Rico, M., Kruse, H., Wlotzka, F., Begemann, F.: Proc. Third Lunar Sci. Conf., Suppl. 3, Geochim. Cosmochim. Acta, Vol. *2*, p. 1251. M. I. T. Press (1972).
13) Wänke, H., Baddenhausen, H., Dreibus, G., Quijano-Rico, M., Palme, H., Spettel, B., Teschke, F.: Multielement analysis of Apollo 16 samples and about the composition of the whole moon. In: Lunar Science – IV ( eds. J. W. Chamberlain and C. Watkins). Houston: Lunar Science Institute (1973).
14) Vinogradov, A. P.: Proc. Second Lunar Sci. Conf., Suppl. *2*, Geochim. Cosmochim. Acta, Vol. *1*, p. 1. M. I. T. Press (1971).
15) Vinogradov, A. P.: Geohimija *7*, 763 (1972).
16) Schonfeld, E., Meyer, C. Jr.: Proc. Third Lunar Sci. Conf., Suppl. 3, Geochim. Cosmochim. Acta, Vol. *2*, p. 1397. M. I. T. Press (1972).

17) Mason, B.: Handbook of elemental abundances in meteorites. New York: Gordon and Breach 1971.
18) Wedepohl, K. H.: Häufigkeit der Elemente in den magmatischen Gesteinen der oberen Erdkruste. In: C. W. Correns: Einführung in die Mineralogie, p. 202. Berlin: Springer 1968.
19) Turekian, K. K., Wedepohl, K. H.: Geol. Soc. Am. Bull. 72, 175 (February 1961).
20) Gast, P. W., Hubbard, N. J., Wiesmann, H.: Proc. Apollo 11 Lunar Science Conf., Suppl. 1, Geochim. Cosmochim. Acta, Vol. 2, p. 1143. Pergamon Press 1970.
21) Hartmann, W. K., Wood, C. A.: The Moon 3, 3 (1971).
22) Howard, K. A., Head, J. W., Swann, G. A.: Proc. Third Lunar Sci. Conf., Suppl. 3, Geochim. Cosmochim. Acta, Vol. 1, p. 1. M. I. T. Press 1972.
23) Turner, G.: Earth Planet. Sci. Lett. 11, 169 (1971).
24) Wood, J. A., Dickey, J. S. Jr., Marvin, U. B., Powell, B. N.: Proc. Apollo 11 Lunar Sci. Conf., Suppl. 1, Geochim. Cosmochim. Acta, Vol. 1, p. 965. Pergamon Press 1970.
25) LSPET (Lunar Sample Preliminary Examination Team): The Apollo 16 Lunar Samples: A petrographic and chemical description of samples from the lunar highlands. Preprint 1972.
26) Shoemaker, E. M., Hait, M. H., Swann, G. A., Schleicher, D. L., Schaber, G. G., Sutton, R. L., Dahlem, D. H., Goddard, E. N., Waters, A. C.: Proc. Apollo 11 Lunar Sci. Conf., Suppl. 1, Geochim. Cosmochim. Acta, Vol. 3, p. 2399. Pergamon Press 1970.
27) Adler, I., Gerard, J., Trombka, J., Schmadebeck, R., Lowman, P., Blodget, H., Yin, L., Eller, E., Lamothe, R., Gorenstein, P., Bjorkholm, P., Harris, B., Gursky, H.: Proc. Third Lunar Sci. Conf., Suppl. 3, Geochim. Cosmochim. Acta, Vol. 3, p. 2157. M. I. T. Press 1972.
28) Hubbard, N. J., Meyer, C., Jr., Gast, P. W.: Earth Planet. Sci. Lett. 10, 341 (1971).
29) Philpotts, J. A., Schnetzler, C. C.: Science 167, 493 (1970).
30) Gast, P. W., Hubbard, N. J.: Science 167, 485 (1970).
31) Schmitt, R. A., Wakita, H., Rey, P.: Science 167, 512 (1970).
32) Keil, K., Prinz, M.: Proc. Second Lunar Sci. Conf., Suppl. 2, Geochim. Cosmochim. Acta, Vol. 1, p. 319. M. I. T. Press 1971.
33) Brown, G. M., Emeleus, C. H., Holland, J. G., Peckett, A., Phillips, R.: Proc. Second Lunar Sci. Conf., Suppl. 2, Geochim. Cosmochim. Acta, Vol. 1, p. 583 M. I. T. Press 1971.
34) Green, D. H., Ringwood, A. E., Ware, N. G., Hibberson, W. O., Major, A., Kiss, E.: Proc. Second Lunar Sci. Conf., Suppl. 2, Geochim. Cosmochim. Acta, Vol. 1, p. 601. M. I. T. Press 1971.
35) Wakita, H., Schmitt, R. A.: Science 170, 969 (1970).
36) Gerling, E. K., Levskii, L. K.: Dokl. Akad. Nauk. SSSR 111, 750 (1956).
37) Zähringer, J., Gentner W.: Z. Naturforsch. 15a, 600 (1960).
38) König, H., Keil, K., Hintenberger, H., Wlotzka, F., Begemann, F.: Z. Naturforsch. 17a, 357 (1962).
39) Signer, P., Suess, H. E.: Earth Sciences and Meteoritics, p. 241. Amsterdam: North Holland Publ. Co. 1963.
40) Hintenberger, H., König, H., Wänke, H.: Z. Naturforsch. 17a, 306 (1962).
41) Hintenberger, H., König, H., Schultz, L., Wänke, H.: Z. Naturforsch. 19a, 327 (1964).
42) Eberhardt, P., Geiss, J., Grögler, N.: Tschermaks Mineral. Petrog. Mitt. 10, 1 (1965).
43) Mazor, E., Anders, E.: Geochim. Cosmochim. Acta 31, 1441 (1967).
44) Lancet, M. S., Anders, E.: Solubilities of noble gases in magnetite: Implications for planetary gases in meteorites. Preprint 1972.
45) Hintenberger, H., Vilcsek, E., Wänke, H.: Z. Naturforsch. 20a, 939 (1965).
46) Wänke, H.: Proc. Intern. Conf. on Cosmic Rays, Jaipur 1963, 473.
47) Wänke, H.: Z. Naturforsch. 20a, 945 (1965).
48) Lunar Sample Preliminary Examination Team: Science 165, 1211 (1969).

49) Hintenberger, H., Weber, H. W., Voshage, H., Wänke, H., Begemann, F., Wlotzka, F.: Proc. Apollo 11 Lunar Sci. Conf., Suppl. 1, Geochim. Cosmochim. Acta, Vol. 2, p. 1269. Pergamon Press 1970.

50) Eberhardt, P., Geiss, J., Graf, H., Grögler, N., Krähenbühl, U., Schwaller, H., Schwarzmüller, J., Stettler, A.: Proc. Apollo 11 Lunar Sci. Conf., Suppl. 1, Geochim. Cosmochim. Acta, Vol. 2, p. 1037. Pergamon Press 1970.

51) Heymann, D., Yaniv, A.: Proc. Apollo 11 Lunar Sci. Conf., Suppl. 1, Geochim. Cosmochim. Acta, Vol. 2, p. 1247 Pergamon Press 1970.

52) Hintenberger, H., Weber, H. W., Takaoka, N.: Proc. Second Lunar Sci. Conf. Suppl. 2, Geochim. Cosmochim. Acta, Vol. 2, p. 1607. M. I. T. Press 1971.

53) Pepin, R. O., Signer, P.: Science *149*, 253 (1965).

54) Epstein, S., Taylor, H. P., Jr.: Proc. Second Lunar Sci. Conf. Suppl. 2, Geochim. Cosmochim. Acta, Vol. 2, p. 1421 M. I. T. Press 1971.

55) Müller, O.: Proc. Third Lunar Sci. Conf., Suppl. 3, Geochim. Cosmochim. Acta, Vol. 2, p. 2059. M. I. T. Press 1972.

56) Eglinton, G.: This volume (1973).

57) Moore, C. B., Lewis, C. F., Larimer, J. W., Delles, F. M., Gooley, R. C., Nichiporuk, W.: Proc. Second Lunar Sci. Conf., Suppl. 2, Geochim. Cosmochim. Acta, Vol. 2, p. 1343. M. I. T. Press 1971.

58) Cameron, A. G. W.: In origin and distribution of the elements (ed. L. H. Ahrens) p. 125. Oxford: Pergamon Press 1968.

59) Heymann, D., Yaniv, A., Adams, J. A. S., Fryer, G. E.: Science *167*, 555 (1970).

60) Heymann, D., Yaniv, A., Walton, J.: Inert Gases in Apollo 14 Fines and the Case of Parentless Ar. In: Lunar Science − III (ed. C. Watkins), p. 376. Houston: Lunar Science Institute 1972, Contr. No. 88.

61) Yaniv, A., Heymann, D.: Proc. Third Lunar Sci. Conf., Suppl. 3, Geochim. Cosmochim. Acta, Vol. 2, p. 1967. M. I. T. Press 1972.

62) Papanastassiou, D. A., Wasserburg, G. J.: Earth Planet. Sci. Lett. *13*, 368 (1972).

63) Morgan, J. W., Ganapathy, R., Laul, J. C., Anders, E.: Lunar Crater Copernicus: Possible Nature of Impact. Preprint 1971.

64) Urey, H. C.: Astrophys. J. *132*, 502 (1960).

65) Anders, E., Ganapathy, R., Keays, R. R., Laul, J. C., Morgan, J. W.: Proc. Second Lunar Sci. Conf., Suppl. 2, Geochim. Cosmochim. Acta, Vol. 2, p. 1021. M. I. T. Press 1971.

66) Laul, J. C., Morgan, J. W., Ganapathy, R., Anders, E.: Proc. Second Lunar Sci. Conf., Suppl. 2, Geochim. Cosmochim. Acta, Vol. 2, p. 1139. M. I. T. Press 1971.

67) Morgan, J. W., Laul, J. C., Krähenbühl, U., Ganapathy, R., Anders, E.: Proc. Third Lunar Sci. Conf., Suppl. 3, Geochim. Cosmochim. Acta, Vol. 2, p. 1377. M. I. T. Press 1972.

68) Anders, E., Ganapathy, R., Krähenbühl, U., Morgan, J. W.: Meteoritic material on the moon. Preprint 1972.

69) Baedecker, P. A., Schaudy, R., Elzie, J. L., Kimberlin, J., Wasson, J. T.: Proc. Second Lunar Sci. Conf., Suppl. 2, Geochim. Cosmochim. Acta, Vol. 2, p. 1037. M. I. T. Press 1971.

70) Baedecker, P. A., Chou, C.-L., Wasson, J. T.: Proc. Third Lunar Sci. Conf., Suppl. 3. Geochim. Cosmochim. Acta, Vol. 2, p. 1343. M. I. T. Press 1972.

71) Wasson, J. T., Baedecker, P. A.: Proc. Apollo 11 Lunar Sci. Conf., Suppl. 1, Geochim. Cosmochim. Acta, Vol. 2, p. 1741. Pergamon Press 1970.

72) Brunfelt, A. O., Heier, K. S.: Proc. Second Lunar Sci. Conf., Suppl. 2, Geochim. Cosmochim. Acta, Vol. 2, p. 1281. M. I. T. Press 1971.

73) Brunfelt, A. O., Heier, K. S., Nilssen, B., Sundvoll, B.: Proc. Third Lunar Sci. Conf., Suppl. 3. Geochim. Cosmochim. Acta, Vol. 2, p. 1133. M. I. T. Press 1972.

74) Wänke, H., Palme, H., Spettel, B., Teschke, F.: Multielement analyses and a comparison of the degree of oxydation of lunar and meteoritic matter. In: The Apollo 15 Lunar Samples (eds. J. W. Chamberlain and C. Watkins) p. 265. Houston: The Lunar Science Institute 1972.

75) Goldstein, J. I. Henderson, E. P., Yakowitz, H.: Proc. Apollo 11 Lunar Science Conf., Suppl. 1, Geochim. Cosmochim. Acta, Vol. *1*, p. 499. Pergamon Press 1970.
76) Goldstein, J. I., Yakowitz, H.: Proc. Second Lunar Sci. Conf., Suppl. 2, Geochim. Cosmochim. Acta, Vol. *1*, p. 177. M. I. T. Press 1971.
77) Goldstein, J. I., Axon, H. J., Yen, C. F.: Proc. Third Lunar Sci. Conf., Suppl. 3, Geochim. Cosmochim. Acta, Vol. *1*, p. 1037. M. I. T. Press 1972.
78) Wänke, H., Wlotzka, F., Jagoutz, E., Begemann, F.: Proc. Apollo 11 Lunar Sci. Conf., Suppl. 1., Geochim. Cosmochim. Acta, Vol. *1*, p. 931. Pergamon Press 1970.
79) Wlotzka, F., Jagoutz, E., Spettel, B., Baddenhausen, H., Balacescu, A., Wänke, H.: Proc. Third Lunar Sci. Conf., Suppl. 3, Geochim. Cosmochim. Acta, Vol. *1*, p. 1077. M. I. T. Press 1972.
80) Wlotzka, F., Spettel, B., Wänke, H.: On the trace element content of metal particles from fines 60601. In: Lunar Science – IV (eds. J. W. Chamberlain and C. Watkins). Houston: Lunar Science Institute 1973.
81) Krähenbühl, U., Morgan, J. W., Ganapathy, R., Anders, E.: Abundance of 17 trace elements in carbonaceous chondrites. Preprint 1972.
82) Oberbeck, V. R., Quaide, W. L.: Icarus *9*, 446 (1968).
83) Papanastassiou, D. A., Wasserburg, G. J., Burnett, D. S.: Earth Planet. Sci. Lett. *8*, 1 (1970).
84) Papanastassiou, D. A., Wasserburg, G. J.: Earth Planet. Sci. Lett. *11*, 37 (1971).
85) Wood, J. A.: Icarus *16*, 462 (1972).
86) Urey, H. C.: The Planets. New Haven: Yale University Press 1952.
87) Baldwin, R. B.: The Face of the Moon. Chicago: University of Chicago Press 1949.
88) Gilbert, G. K.: Bull. Phil. Soc. Wash. *12*, 241 (1893).
89) Eggleton, R. E.: Thickness of the Apenninian series in the Lansberg region of the Moon. Astrogeol. Studies. Ann. Rept. Aug. 1961 – Aug. 1962. Pt. A. U. S. Geol. Survey open – File Rept., p. 19 (1963).
90) Eggleton, R. E.: Preliminary geology of the Riphaeus Quadrangle of the Moon and definition of the Fra Mauro Formation. Ann. Prog. Rept., Aug. 1962 – July 1963. Pt. A. U. S. Geol. Survey open – file Rept., p. 46 (1964).
91) Dence, M. R., Plant, A. G.: Proc. Third Lunar Sci. Conf., Suppl. 3, Geochim. Cosmochim. Acta, Vol. *1*, p. 379. M. I. T. Press 1972.
92) Engelhardt, W. von, Arndt, J., Stöffler, D., Schneider, H.: Proc. Third Lunar Sci. Conf., Suppl. 3, Geochim. Cosmochim. Acta, Vol. *1*, p. 753. M. I. T. Press 1972.
93) Papanastassiou, D. A., Wasserburg, G. J.: Earth Planet. Sci. Lett. *12*, 36 (1971).
94) Husain, L. Schaeffer, O. A., Funkhouser, J., Sutter, J.: Proc. Third Lunar Sci. Conf., Suppl. 3, Geochim. Cosmochim. Acta, Vol. *2*, p. 1557. M. I. T. Press 1972.
95) Turner, G., Huneke, J. C., Podosek, F. A., Wasserburg, G. J.: Proc. Third Lunar Sci. Conf., Suppl. 3, Geochim. Cosmochim. Acta, Vol. *2*, p. 1589 M. I. T. Press 1972.
96) Silver, L. T.: U–Th–Pb Isotope systems in Apollo 11 and 12 regolithic materials and a possible age for the Copernicus impact event. Paper presented at Amer. Geophys. Union 52nd Ann. Meeting. Washington, D. C. 1971.
97) Eberhardt, P., Eugster, P., Geiss, J., Grögler, N., Schwarzmüller, J., Stettler, A., Weber, L.: When was Apollo 12 KREEP ejected? In: Lunar Science – III (ed. C. Watkins), p. 206. Houston: Lunar Science Institute Contr. No. 88. 1972.
98) Wasson, J. T., Baedecker, P. A.: Proc. Third Lunar Sci. Conf., Suppl. 3, Geochim. Cosmochim. Acta, Vol. *2*, p. 1315. M. I. T. Press 1972.
99) Hartmann, W. K.: Lunar crater counts. VI: The young craters Tycho, Aristarchus and Copernicus. Comm. Lunar Planet. Lab. *7*, 145 (1968).
100) Laul, J. C., Wakita, H., Showalter, D. L., Boynton, W. V., Schmitt, R. A.: Proc. Third Lunar Sci. Conf., Suppl. 3, Geochim. Cosmochim. Acta, Vol. *2*, p. 1181. M. I. T. Press 1972.
101) Wood, J. A.: Icarus *2*, 152 (1963).
102) Anders, E.: Space Sci. Rev. *3*, 583 (1964).
103) Grossman, L.: Geochim. Cosmochim. Acta *36*, 597 (1972).

104) Lewis, J. S.: Earth Planet. Sci. Lett. *15*, 286 (1972).
105) Urey, H. C.: Geochim. Cosmochim. Acta *2*, 269 (1952).
106) Urey, H. C.: Ap. J. Suppl. *6*, 147 (1954).
107) Lord, H. C.: Icarus *4*, 279 (1965).
108) Larimer, J. W.: Geochim. Cosmochim. Acta *31*, 1215 (1967).
109) Anders, E.: Accounts Chem. Res. *1*, 289 (1968).
110) Anders, E.: Ann. Rev. Astronomy Astrophys., Vol. *9*, 1 (1971).
111) Cameron, A. G. W.: Physical conditions in the primitive solar nebula. In: Meteorite Research. (ed. P. M. Millman). Reidel: Dordrecht 1969.
112) Grossman, L.: Condensation, chondrites and planets. Thesis. New Haven: Yale University 1972.
113) Wänke, H., Baddenhausen, H., Palme, H., Spettel, B.: On the chemistry of the Allende inclusions and their origin as high temperature condensates. To be published.
114) Langseth, M. G., Jr., Clark, S. P., Chute, J., Jr., Keihm, S.: The Apollo 15 Lunar Heat Flow Measurement. In: Lunar Science – III (ed. C. Watkins), p. 475, Houston: Lunar Science Institute 1972 Contr. No. 88.
115) Gast, P. W.: Science *147*, 858 (1965).
116) Gast, P. W.: J. Geophys. Res. *65*, 1287 (1960).
117) Wänke, H.: Z. Naturforsch. *16a*, 127 (1961).
118) Gast, P. W.: Paper presented at CNRS Colloquium "On the Origin of the Solar System". Nice, France, April 1972.
119) Anderson, D. L.: IAGC Symp. Cosmochem. Cambridge, Mass., August 1972, preprint.
120) Sonett, C. P., Smith, B. F., Colburn, D. S., Schubert, G., Schwartz, K.: Proc. Third Lunar Sci. Conf., Suppl. 3, Geochim. Cosmochim. Acta, Vol. *3*, p. 2309. M. I. T. Press 1972.
121) Albee, A. L., Chodos, A. A., Gancarz, A. J., Haines, E. L., Papanastassiou, D. A., Ray, L., Tera, F., Wasserburg, G. J., Wen, T.: Earth Planet. Sci. Lett. *13*, 353 (1972).
122) Tera, F., Eugster, O., Burnett, D. S., Wasserburg, G. J.: Proc. Apollo 11 Lunar Sci. Conf., Suppl. 1, Geochim. Cosmochim. Acta, Vol. *2*, p. 1637. Pergamon Press 1970.
123) Papanastassiou, D. A., Wasserburg, G. J.: Earth Planet. Sci. Lett. *17*, 52 (1972).
124) Latham, G., Ewing, M., Dorman, J., Lammlein, D., Press, F., Toksöz, M. N., Sutton, G., Duennebier, F., Nakamura, Y.: Proc. Third Lunar Sci. Conf., Suppl. 3, Geochim. Cosmochim. Acta, Vol. *3*, p. 2519. M. I. T. Press 1972.
125) Reynolds, R. T., Fricker, P. E., Summer, A. L.: J. Geophys. Res. *71*, 573 (1966).
126) Toksöz, M. N, Press, F., Dainty, A., Anderson, K., Latham, G., Ewing, M., Dorman, J., Lammlein, D., Sutton, G., Duennebier, F.: Proc. Third Lunar Sci. Conf., Suppl. 3, Geochim. Cosmochim. Acta, Vol. *3*, p. 2527. M. I. T. Press 1972.
127) Kaula, W. M.: Science *166*, 1581 (1970).
128) Ringwood, A. E., Essene, E.: Proc. Apollo 11 Lunar Sci. Conf., Suppl. 1, Geochim. Cosmochim. Acta, Vol. *1*, p. 769. Pergamon Press 1970.
129) Gast, P. W. Giuli, R. T.: Earth Planet. Sci. Lett. *16*, 299 (1972).
130) Ringwood, A. E.: Earth Planet. Sci. Lett. *8*, 131 (1970).
131) Ash, M. E., Shapiro, I. I., Smith, W. B.: Science *174*, 551 (1971).
132) Cameron, A. G. W.: Nature *240*, 299 (1972).

Received January 29, 1973

# Advances in Inorganic Geochemistry

**Dr. Harald Puchelt**
Mineralogical Institute, Departement of Geochemistry, University of Tübingen

## Contents

| | | |
|---|---|---|
| 1. | Introduction | 156 |
| 2. | Analytical Geochemistry (Data) | 156 |
| 3. | Analytical Geochemistry (Methods) | 157 |
| 4. | Statistical Methods and Evaluations | 158 |
| 5. | Ore Deposits | 159 |
| 6. | Geochemical Prospecting | 160 |
| 7. | Hydrogeochemistry | 160 |
| 8. | Chemistry of the Oceans: Development and Present State | 161 |
| 9. | Geochemistry of Stable and Radioactive Isotopes | 162 |
| 10. | Environmental Geochemistry | 165 |
| 11. | Crystal Chemistry and Element Partitioning | 166 |
| 12. | Geochemical Models | 167 |
| 13. | Literature | 168 |
| 14. | References | 169 |

## 1. Introduction

Correns (1969) has given a vivid description of the development of ideas concerning the elemental composition of the earth's crust and the discovery of the elements. Only 150 years ago (1821), the term "chemistry of the earth" was used by Berzelius, as synonymous with the term "mineralogy". In the middle of the last century (1845—1854), G. Bischof published his three-volume work "Lehrbuch der chemischen und physikalischen Geologie". Chemists (like Bunsen) and earth scientists became aware of the potential of a chemical approach to the problems of earth and hydrosphere. Clarke published the first edition of his "Data of Geochemistry" in 1908 and the fifth edition (1924) remained a standard reference for many years. Vernadsky (1863—1945) and Fersman (1883—1945) were the initiators of geochemical thinking in Russia.

The essentials of geochemistry, the laws of distribution of the elements, and crystal chemistry were outlined and investigated to a great extent by V. M. Goldschmidt and his students during his years in Oslo and Göttingen. His posthumous book "Geochemistry" (1954) was a landmark in the development of this field. Another early textbook of geochemistry, but essentially promoting the ideas of Goldschmidt, was that of Rankama and Sahama (1950). In the last 25 years geochemistry has developed very fast and diversified into a number of branches which already seem to have taken on a life of their own.

This short paper surveys some of the topics which seem to the author to be of major importance in the development of this field.

## 2. Analytical Geochemistry (Data)

The recent state of analytical geochemistry is documented by the chapters of the Handbook of Geochemistry (Wedepohl ed. from 1969 onward) on the individual elements.

A Russian approach is the book by Vlasov (1964) which also describes the individual elements.

The "Data of Geochemistry" 6th edition (Prof. Paper U.S. G.S. 440 A-JJ) tackles the geochemistry of domains such as atmosphere, biosphere, rock-forming minerals, igneous rocks, shales, sandstones, carbonate rocks, evaporites or deposits, etc.

The following is a selection of recent reviews:

Graf (1960): Geochemistry of carbonates,
Goldberg (1961): Geochemistry of deep-sea sediments,
Taylor (1968): Geochemistry of andesites,
Garrels, R. M., Mackenzie, F. T. (1971): Geochemical approach to evolution of sedimentary rocks,
Manson (1967): Geochemistry of basalts (major elements),
Prinz (1967): Geochemistry of basalts (trace elements),
Degens (1965): Geochemistry of sediments.

The literature is surveyed in:
  Mineralogical Abstracts,
  Chemical Abstracts,
  Zentralblatt für Mineralogie,
  Bulletin Signalitique.

## 3. Analytical Geochemistry (Methods)

*Analytical methods:* Two developments have greatly influenced analytical geochemistry: sophisticated analytical methods and the technique of comparing results against worldwide geochemical standards.

Inorganic geochemistry is one of the few fields where the old "wet chemistry" of silicate, phosphate, carbonate and sulfide materials has been retained and even developed further.

*Spark and arc spectroscopy* was developed by means of the automatically recording Quantometers into a routine method with simultaneous immediate data output for several elements (Danielson *et al.*, 1959). The preparation of all sorts of materials has been automated. Matrix effects (inhomogeneities and interferences) were minimized by dissolving the substances and subjecting them to ion exchange on an epoxy resin which is then continuously fed into the arc (Danielson, 1962). The data are monitored by photocells set on the element lines behind a high-resolution grating. The data are processed by a small computer, the output being given directly in weight % of the oxide.

While normal spark spectroscopy mostly uses powdered samples and consequently gives averages for a comparatively large sample, laser excitation allows to analyze the composition of particles of 10 $\mu$ diameter.

Yet conventional spark and arc spectroscopy is not outdated since it allows the recording and thus the detection of all excitable elements in the sample at a single exposure, an advantage very few other methods can offer.

*X-ray spectroscopy* is nowadays applied mostly in the form of X-ray fluorescence, where scanning monochannel machines, sequence spectrometers and simultaneous spectrometers are used in wavelength dispersive X-ray fluorescence. The introduction of bent analyser-crystals extended the method to smaller samples, thus marking another step toward microprobe analysis.

Crystals with a high dispersion allow the various oxidation states of an element to be distinguished according to the observed peak shift (FeO, $Fe_3O_4$, $Fe_2O_3$, $Ti^{3+}$, $Ti^{4+}$, Pavicevic *et al.*, 1972).

A rather new development in this field is energy dispersive X-ray fluorescence using energy-dispersive Si detectors with attached multichannel analysers. X-irradiation excited by radio-isotopes makes this method independent of large X-ray tubes. The application of nondispersive X-ray fluorescence is somewhat restricted by its limit of resolution.

The electron microprobe represents the greatest advance in analysing areas down to 1 $\mu^2$. This instrument is most valuable for studying terrestrial and lunar rocks and minerals.

*Atomic absorption:* Atomic absorption spectrometers are widely used in geochemistry and most elements from major constituents down to traces can be determined by this means (Angino and Billings, 1972).

AAS is widely used in geochemical prospection where a special accessory, the graphite cell, was recently introduced. Elements down to $10^{-13}$ g can be determined by this accessory. A less widely applied technique is atomic fluorescence, which gives especially good results for certain elements.

*Neutron activation* is much used for main and trace element determinations in geological samples (Brunfelt and Steinnes, 1971). Thanks to the accessibility of thermal neutrons from nuclear reactors and fast (14 MeV) neutrons from commercial tube sources and the availability of high-resolution Ge(Li) detectors and multichannel analysers, many laboratories use this method. The resolving power of the detectors has been much improved from the old NaJ(Tl) crystals to the modern semiconductor crystals, so that the method is now in most cases purely instrumental and no radiochemical separation is necessary. How many nuclides can be determined depends on the half-lives of the nuclides generated, i.e. on the distance of the analysing system from the radiation source. The neutron tube has been improved to yield fluxes of $10^{11}$ n/sec and is today applied to determinations of major and trace elements and to elements like oxygen and fluorine that are difficult to analyze. The accessibility of Cf-252 has made neutron activation a field method which can be applied in situ for the surveying of rock sequences in bore holes (Senftle, 1972).

The measurement of natural gamma rays from rocks and minerals requires basically the same equipment as $\gamma$-ray spectrometry in neutron activation, the only difference being that large crystal or semiconductor detectors in a more sophisticated arrangement are necessary to record the low natural activity (Adams and Gasparini, 1970).

*Mass spectroscopy:* Spark-source mass spectroscopy has been applied to certain terrestrial and lunar materials with excellent results because of its very low detection limit and the exactness of the results. Drawbacks are the cost of the instruments and the time needed for each analysis.

Gas mass spectroscopy is generally applied to H, D, $^{16}$O, $^{18}$O, $^{14}$N, $^{15}$N, $^{12}$C, $^{13}$C, $^{32}$S, $^{34}$S.

Solid-state mass spectroscopy is much used in geochronology, but also in Sr, Rb, Pb, and U isotope investigations.

## 4. Statistical Methods and Evaluations

The analytical data are so voluminous as to require the application of statistical methods, particularly when planning geochemical investigations, e.g. geochemical prospection (cf. Hawkes and Webb, 1962), sampling, sample preparation (crushing, splitting, grinding), analysis (Shaw and Bankier, 1954; Thiergärtner, 1968).

The importance of representative sampling is now recognized; the magnitude of errors in analytical procedures was demonstrated for the first time by Fairbairn *et al.* (1951) on the basis of a cooperative investigation of chemical and spectrochemical silicate rock analyses. Things have improved since then, and objective parameters, like "standard deviation" and coefficient of variance have been established. A further very important application of statistics is to determine distribution laws (mostly lognormal) for elements (Gaddum, 1945; Ahrens, 1954 a, b, 1957, 1963 a, b; Shaw, 1961). Bryan *et al.* (1969) estimated proportions in petrographic mixing equations in order to evaluate fractional crystallization, assimilation and mixing of magmas and relate them to variations in igneous rock series.

In geochemical prospecting "trend surface analysis" used with an appropriate computer program (Kulms and Siemes, 1970; Nichol, 1969; Nordeng, 1965) can plot isoelement maps from the analytical data. Background information for expected values and anomalies in prospecting were obtained by Bölviken (1971).

Mathematical treatment of geochemical data cannot, of course, improve their quality but does ensure proper interpretation of the data and the procedures used to obtain them. Multivariate analysis can bring a mass of data into manageable form and reveal relationships between variables which graphical analysis would not (Middleton, 1963).

A brief general introduction to data evaluation with a discussion of analytical errors, and statistical procedures was published by Shaw (1969).

## 5. Ore Deposits

Geochemical information has added much to the understanding of the genesis of economic deposits. It covers chemistry of ore minerals in the broadest sense, liquid inclusions (Roedder, 1972), the conditions in the surrounding country rocks and laboratory work with theoretical evaluation (Barnes, 1967). Hydrothermal syntheses of ore and gangue minerals (cf. Holland, 1973) indicated conditions of genesis and stability of the ores. The boundaries of many stable phases have been determined in relation to temperature, pressure, composition, and fugacity of the gases involved. Experiments (cf. Barton and Skinner, 1967; Kullerud, 1970; Helgeson, 1970) have shown which ore-forming sulfides form by cooling of the respective melts and which display characteristic of incongruent melting. The parameters of the ore-bearing fluids are characterized by either chloride, carbonate or sulfide assemblages. Barnes (1972) studied the conditions in which precipitation from solutions could occur: oxidation, increasing acidity, decreasing temperature, dilution and decreasing pressure for sulfide complexes; mixing with sulfide-rich solutions or rocks, reduction, decreasing acidity, temperature and pressure in the case of chloride complexes. He was then able to group certain types of deposits according to the conditions of their deposition, obtained by analysis of paragenetic assemblages, liquid inclusion data, stable isotope ratios and trace element distribution. Helgeson (1972) determined thermodynamic constraints on mass transfer in hydrothermal systems and pointed out that in many cases deposition of ores and minerals is due to kinetic processes rather than adjusting to equilibrium conditions.

159

## 6. Geochemical Prospecting

Since the 1930s chemical methods have been intensively applied to prospection in Russia, where this discipline is called "metallometry" (Fersman, 1939; Vinogradov, 1954). Every metal was searched for by metallogists all over the Eurasian area of the country. During and after World War II much energy was put into mineral prospection, mainly by geochemical methods: trace analyses, soil investigations, vegetation surveys, rock surveys and drainage investigations. The principles of geochemical prospecting were set out by Hawkes and Webb (1962), who also listed those government agencies active in this field.

Nuclear techniques are dealt with in several works (IAEA Panel 1969, 1971). Those techniques have been developed further so that today probes using radioactive sources are available which provide activation analyses (Senftle, 1972; Vogg, 1973).

In 1970, the Third International Geochemical Exploration Symposium was held in Toronto, Canada: 110 papers reported on activities in this field all over the world (Boyle and Mc Gerrigle, 1971).

The 24th International Geological Congress in Montreal in 1972 discussed exploration geochemistry in glaciated terrains (Boyle and Shaw, 1972).

A new journal, Journal of Geochemical Exploration, was launched by the Association of Exploration Geochemists in 1972.

## 7. Hydrogeochemistry

Hydrogeochemistry deals with precipitation, surface and ground waters as well as with sea water and pore solutions in sediments. The physicochemistry of these aqueous systems is presented in a lucid manner by Garrels and Christ (1965) and Stumm and Morgan (1970). The chemical composition of waters is the result of interaction with solid matter (chemical weathering, ion exchange, dissolution and precipitation), liquids (oil) or gases (rain exchanging with air, $CO_2$ and $H_2S$ with pore solutions) and of biological activity. Surface water geochemistry is important both for evaluating geochemical cycles and for geochemical prospection, also for studies of suitability as drinking water. Attempts to compile worldwide composition data were made by Durum *et al.* (1960) and Livingstone (1963). Today environmental influences are more often studied than the natural geochemistry of rivers and lakes.

Pore solutions are present in practically all sediments which are not completely consolidated and often contain more dissolved solids than sea water. If the aqueous phase was sea water when the sediment was formed, its composition was in most cases changed by physical, chemical-mineralogical, and biological processes. Pore solutions provide the chemical components for diagenetic alterations.

Schoeller (1955) discussed the role of subsurface waters in oil deposits.

The geochemistry of subsurface waters was reviewed up to 1963 by White *et al.* (1963); it was the subject of a special issue of Chemical Geology (Angino and Billings eds., 1968). The role of pore solution in sediment diagenesis has been thoroughly discussed by von Engelhardt (1973) and the metal content of geothermal fluids and mineral waters by Ellis (1967) and White (1967).

## 8. Chemistry of the Oceans: Development and Present State

Rubey (1950) formulated questions concerning the history of the ocean, and Holland (1972) has attempted to solve the problem of ocean development, which is fundamental. Conway (1942, 1943, 1945) discussed whether the ocean had from its very beginning in the Precambrian the same volume and composition as today. Holland (1972) examined the geological records provided by evaporite deposits and affirmed that the composition of the ocean never was fundamentally different. Holland's calculations set definite limits for variations in pH, $CO_2$ pressure and related $CO_3^=$ and $HCO_3^-$ concentration, and $Mg^{++}$, $Ca^{++}$ and sulfate concentrations and ratios to each other, but do not give the total amount of sea water present at each geological age.

Using the classic calculations of Goldschmidt (1954) on the weathering cycle, Rubey (1950) pointed out the excess of several ions in sea water compared to concentrations derived from weathering of igneous rocks. The continuous addition of those elements through volcanic activity is still observed today. In respect to the water content of volcanic exhalations, the problem of recycling is not yet solved, nor is it clear whether some of the iron and manganese on the seafloor, obviously of volcanic origin, is recycled (Boström, 1967).

An important step forward in understanding elemental abundance in the ocean was the calculation of residence times for each element (Barth, 1952). Some elements, like Be and Al, have a residence time of the order of $10^2$ years, while the value for Na is $10^8$ years. Any large-scale change in the chemistry of the oceans would have had to influence the input and output of the main ions over at least millions or even billions of years.

Krauskopf (1956) and Goldberg (1965) discussed the factors responsible for the ions present in the oceans of today, especially biologic activity, since life processes are intimately involved in ocean chemistry. Holland (1972) gives emphasis to their kinetics.

The gross homogeneity of the oceans was recognized by Dittmar (1884) who analysed 74 water samples from all the oceans. Though his values are still largely valid, there are some differences in salinity ascribed to the movement of water masses (Dietrich, 1963), itself due to differences in density. Craig and Gordon (1965) studied the mixing of these water bodies. A box model for oceanic mixing was calculated by Broecker et al. (1961), using carbon-14.

Distribution profiles of anions were established by Sverdrup *et al.* as early as 1942, and many trace constituents were analysed by Turekian and coworkers (1965a/b, 1966a/b).

Important advances in the understanding of ocean chemistry, transport mechanisms, equilibria and the kinetics involved, were stimulated by the GEOSECS program. The aims of the program and the work of the participating nations are briefly outlined by Craig (1972). The GEochemical Ocean SECtions sampled, measured and analysed in connection with the International Decade of Ocean Exploration were chosen for the study of the circulation and mixing of the world oceans. A collection of results for the period 1970–1971 was published in Earth and Planetary Science Letters *16* (1972) 47–145.

Closely related to the behavior of trace elements in seawater is the composition of pelagic sediments, also covered by some parts of the program.

The effects of pressure on equilibria in the oceans within depth profiles have been studied mostly in relation to the problem of calcium carbonate saturation in this environment (Millero, 1969; Berner, 1965; Millero and Berner, 1972; Edmond and Gieskes, 1970). The early calculations by Owen and Brinkley (1941) concerning the effect of pressure upon ionic equilibria in salt solutions have been extended to studies of $BaSO_4$ solubility at different depths (Chow and Goldberg, 1960) and to the pressure dependence of sulfate associations (Fisher, 1972).

The connection between weathering, evaporation, bacterial activity in marine sediments in shallow water areas and stable sulfur isotope distribution was studied by Holser and Kaplan (1966), Nielsen and Ricke (1969), Nielsen (1968), and Eremenko and Pankina (1972). The sulfur isotope ratios of evaporite sulfates show enrichment of $^{34}S$ in the Devonian deposits ($\delta \approx$ +23 $^o$/oo), a pronounced "dip" in the Permian ($\phi$ $\delta \approx$ +12 $^o$/oo), values of +20 $^o$/oo in the Triassic, a fall (+17 to +15 $^o$/oo) in the Jurassic and Cretaceous, stabilizing in the Tertiary to the Recent value of +20 $^o$/oo.

The high $\delta^{34}S$ values of some Precambrian evaporites and the fluctuating values of certain Precambrian sulfides are thought to show the presence of free oxygen and bacteria in Proterozoic times, because at low temperatures only bacterial fractionation – an indication of early organic life – could have produced such differences in $^{34}S$ content of sulfates and sulfides.

## 9. Geochemistry of Stable and Radioactive Isotopes

If we assume that the isotopic composition of the elements was set by the original processes of nucleosynthesis, deviations from the primordial ratios can originate by:
1. fractionation of light stable and radioactive isotopes induced by equilibrium shifts between different bonding or oxidation stages, or by kinetic effects;
2. generation of new stable isotopes through radioactive decay;
3. generation of new isotopes by fission of uranium.

Isotope investigations are used for age determinations, for studying the mode, source and temperature of formation of rocks, minerals and deposits, for determining paleotemperature, and for establishing kinetic models of water cycles in hydrosphere and atmosphere.

This whole field was reviewed by Rankama (1954, 1963). Craig *et al.* (1964) collected many isotope papers from the school of Urey. Isotope geochemistry has grown so fast that today a single scientist can hardly survey all the isotopes, methods and applications. A brief introduction to stable isotope geochemistry, its progress, and its possibilities was recently published by Hoefs (1973).

*Age Determinations*

Radioactivity was used for dating before mass spectrometric techniques were known and applied to the problem, but real progress in this field had to await exact determinations of both parent *and* daughter isotopes. Most of the interfering factors are now known so that not only can the formation of a rock be dated, but also the metamorphism and metasomatism which influenced the rock later on (cf. Jäger, 1970). Thus a wide range of ages can be dated with the appropriate radioisotopes. Of the many general books on geochronology, only York and Farquhar (1972) and Hamilton (1965) shall be mentioned here. The abstracts of contributions to the 1st and 2nd European Colloquia of Geochronology (1971 and 1972) provide a recent survey of progress.

| Radioactive isotope | Daughter | Halflife (years) | Monograph |
|---|---|---|---|
| T | $^3$He | 12.4 | Libby, 1959 |
| | | | Chatters, 1965 |
| $^{14}$C | $^{14}$N | $5.73 \cdot 10^3$ | Grey, 1973 |
| $^{40}$K | $^{40}$Ar | $1.33 \cdot 10^9$ | Schaeffer und Zähringer, 1966 Dalrymple and Lanphere, 1969 |
| $^{87}$Rb | $^{87}$Sr | $5.0 \cdot 10^{10}$ | Faure and Powell, 1972 |
| $^{187}$Re | $^{187}$Os | $\approx 5 \cdot 10^{10}$ | Hirt *et al.*, 1963 |
| $^{232}$Th | $^{208}$Pb | $1.39 \cdot 10^{10}$ | |
| $^{235}$U | $^{207}$Pb | $7.1 \cdot 10^8$ | Doe, 1970 |
| $^{238}$U | $^{206}$Pb | $4.5 \cdot 10^9$ | |

*Stable isotope fractionation* was predicted and calculated by Urey and coworkers as early as 1932. Bigeleisen (1961) dealt with the statistical mechanics of isotope effects. Hydrogen isotope fractionation is mainly connected with the processes in the hydrosphere, but its equilibrium and kinetic effects were also studied for water of crystallization, ore-forming fluids, as represented by liquid inclusions, and biological cycles.

Carbon isotopes are fractionated in biological cycles and in inorganic oxidation-reduction reactions. In equilibrium systems of various oxidation states, $^{13}$C enrichment up to 500 °K occurs in the following sequence:

$$CO < CH_4 < C_D < CO_2 < CO_3^{2-}.$$

The materials with highest $^{13}$C concentrations are several carbonates from carbonaceous chondrites. Carbonates from the cap rock of salt domes, formed bacterially from $CaSO_4$ and oxidized organic substances, show strong $^{13}$C

H. Puchelt

depletion (down to $-44$ $^0/oo$ vs. PDB standard). Normal marine limestone and dolomite has values between +4 and $-6$ $^0/oo$ vs PDB standard. A bibliography of papers on carbon isotopes was published by Weber (1966).

In the range $0-1000$ °C oxygen isotopes are extensively applied as temperature indicators in problems connected with formation conditions of solid and liquid phases. In the low-temperature range oxygen isotope analyses are mostly applied to typical sedimentary minerals. If one assumes equilibrium between seawater and the mineral (mostly carbonate) which formed from it and no subsequent diagenetic disturbing, temperature differences down to 0.5 °C can be determined with the available techniques. Paleotemperature analysis including phosphate geothermometry is discussed by Bowen (1966).

Oxygen thermometry for minerals in igneous and metamorphic rocks was surveyed by Epstein and Taylor (1967). These authors reviewed older analytical and experimental work and showed how useful many equilibrium silicate and oxide pairs (quartz-muscovite, quartz-magnetite, quartz-garnet among many others) can be for recalculation of equilibration temperatures.

Sulphur isotopes ($^{32, 33, 34, 36}S$) fractionate strongly in the earth's crust because (1) the element occurs in different oxidation states with differential preference for heavy isotopes, (2) the existence of volatile and easily soluble compounds favors kinetic separations, and (3) it is involved in biogenic cycles where the oxidation state is easily changed and kinetic processes are important. From theoretical calculations of Bigeleisen (1961) and data on the isotopic properties of sulphur compounds by Sakai (1957, 1968), the amount of S isotope fractionation and its temperature dependence is known. The information on experimental inorganic isotope fractionation in coexisting sulphide minerals which occur naturally was summarized by Thode (1970), who also discussed the application of S isotopes from sulphides for geothermometry (cf. also Sakai, 1971). Analytical work on all types of sulphur compounds which occur in nature has been reviewed by Nielsen (1973).

| Isotope pairs | Monographs or reviews |
|---|---|
| H/D | H. D. Taylor, 1973 |
| $^{12}C/^{13}C$ | Schwarcz, 1968 |
| $^{14}N/^{15}N$ | Wlotzka, 1972 |
| $^{16}O/^{18}O$ | H. P. Taylor, 1973 |
| | Bowen, 1966 (Paleotemperature anal.) |
| | Thode, 1970 (exper. isotope distr.) |
| $^{32}S/^{34}S$ | Nielsen, 1973 |
| $^{86}Sr/^{87}Sr$ | Faure and Powell, 1972 |
| $^{204}Pb/^{206}Pb$ | Doe, 1970 |

Bacterial reduction of sulfate in an anaerobic environment with large isotope fractionation between the residual sulphate and the sulphide is the source of sulphur and sulphide in many deposits. The recognition that many sulphur

164

deposits are the product of bacterial reduction of sulphates was one of the great achievements of isotope geochemistry.

Jensen (1962) has reviewed the part bacteria play in S isotope fractionation.

Changes in the isotope composition of marine sulphates during geological time is another exiting topic of S isotope geochemistry (Holser and Kaplan, 1966; Nielsen, 1970; Vinogradov, 1972).

Strontium isotopes, though not fractionated easily in nature, are a valuable tool for genetic investigations on rocks and minerals beside their application for dating by the Rb/Sr method. The distribution of elements in the different parts of the earth's crust and mantle depends on their crystal chemistry. Thus, Sr is retained mostly in the mantle, while Rb preferentially accumulates in the crust. The ratio $^{87}Sr/^{86}Sr$ is continuously and very slowly rising in the mantle but may be strongly increased in the crust. Sr isotope ratios of 0.701–0.703, being indicative of mantle or deep-crust material, have clarified many petrogenetic problems. Investigations in this field, especially on the basalt problem, were carried out by Gast (cf. 1967).

Recently indications were found that natural fission took place more than $10^8$ years ago in a uranium deposit in Gabon/Africa (Bodu et al., 1972). Low $^{235}U/^{238}U$ ratios (down to 0.440% $^{235}U$) and anomalous Ce, Sm, and Eu isotope compositions are evidence of this unusual event. The normal $^{235}U$ content in natural uranium is 0.720%, but 1.7 billion years ago the concentration was about 3%, which is much closer to a critical assemblage (Neuilly et al., 1972). This is the first reported example of a natural nuclear reactor. P. K. Kuroda in 1956, using data of nuclear physics, calculated that such an event could have happened in geological time, approximately two billion years ago.

High neutron fluxes must also be the reason for the naturally occurring isotope $^{147}Pm$ (half life 2.7 years) found by Meier et al. (1972).

Natural radioactive isotopes produced by cosmic rays ($^{14}C$, $^{32}Si$, $^{226}Re$) are used in determining the mixing and circulation of ocean water (Kharkar et al., 1963; Somayajulli et al., 1973; Broecker et al., 1967; Lal, 1966; Craig, 1969), while $^3H$ and $^{14}C$ have extensive applications in hydrogeology.

## 10. Environmental Geochemistry

Many human activities interfere with the normal conditions of the upper crust, the hydrosphere and the atmosphere. Geochemistry can monitor and possibly direct the development of more environment-conscious attitudes. For decades geochemists have been busy collecting data on sediment cover, soil, and the hydrosphere, atmosphere and biosphere of our planet. Early observations demonstrated an almost perfect equilibrium system; today atmosphere, hydrosphere, and soil in many places are far out of balance (cf. Banar, Förstner and Müller, 1972 a, b, a report on river waters and sediments in Germany with a good reference list). All biological activity is seriously affected (cf. Müller and Förstner, 1973), not only in freshwater systems but also in parts of the oceans, where disequilibrium conditions have been observed locally. The German Science

Foundation has started a program to identify the problems caused by dangerous substances in water and to recommend remedial measures.

Geochemistry can in many cases be applied to determine limits of tolerance and give warnings of dangerous increases or trace element deficiencies in water and plants. The laboratories are equipped with sophisticated instruments, which can detect all sorts of pollution, even in the atmosphere (Steinberg, 1971).

A direct link with medical applications is demonstrated by a Memoir of the Geological Society of America on "Environmental Geochemistry in Health and Disease" (Cannon and Hopps, 1971): many routine techniques used in geochemical prospecting can with slight modifications be extended to the study of geomedical problems.

This important field is now represented by the International Society for Environmental Geochemistry and Health. "Trace Substances in Environmental Health" (Hemphill, 1972) is a report of its first conference in Missouri.

In environmental fields geochemists have to cross the frontiers of neighbouring disciplines, but they are dealing with fundamental data for which they have a well-tested analytical apparatus.

## 11. Crystal Chemistry and Element Partitioning

The rules of crystal chemistry as outlined by Goldschmidt (1937) led to many insights into the behavior and distribution of elements during rock or mineral formation from melts and solutions. Many aspects of crystal chemistry — geometrical framework of crystal structures, effective radii of atoms and ions, chemical bonds, crystal chemistry of real crystals, diffusion and solid-state reactions — were treated by Zemann (1969). A redetermination of ionic radii (Shannon and Prewitt, 1969) very much improved the applicability of this concept, enabling differences in coordination number and structural sites to be taken into consideration. The application of crystal field theory to the geochemistry of transition elements provides better information than bond energy in crystal phases alone (Burns and Fyfe, 1967). On the basis of site-preference energy parameters good predictions can now be made concerning the uptake of transition elements from the magma into crystals.

The partitioning of elements between aqueous solution and crystals and between silicate melts and crystalline solids in the light of thermodynamic theory was excellently reviewed by McIntyre (1963).

Two approaches have been adopted to find the laws of distribution:
1. analyses of coexisting minerals in volcanic rocks and their glass base;
2. experimental investigations on the distribution of elements between solutions (including hydrothermal systems), melts or subsolidus systems and the crystals growing from them.

Several purposes underlie this work: one is to establish general laws of distribution for certain main and trace elements; another is to correlate trace element partitioning with temperature of formation or equilibration (Häkli and Wright, 1967; Kudo and Weil, 1970); a third is to acquire information on the partial pressure of oxygen during mineral formation from $Eu^{2+}$ to $Eu^{3+}$ ratios (Philpotts, 1970), or to speculate on the genesis of basalt (Philpotts, Schnetzler, 1970) or the sequence of deposition in salt deposits (Braitsch, 1963; Schock and Puchelt, 1971).

This field is surveyed by Philpotts and Schnetzler (1972) who also presented their own data on the range of partitioning for K, Rb, Ba, Sr and many rare earths between rock-forming minerals (phenocrysts) and matrix. The thermo-dynamics and kinetics of chemical exchange reactions in crystalline rocks were discussed by Kretz (1972).

Element distribution in feldspars was studied by Jasmund and Seck (1972) (subsolidus synthesis) and Iiyama (1972) (hydrothermally).

Dale and Henderson (1972) investigated trace element distribution in phenocrysts (olivine, pyroxene) and basalt matrix, with much consideration of crystal field effects. Banno and Matsui (1973) published a new formulation of partition coefficients for trace element distribution between minerals and magmas, based on the concept of chemical potentials for liquids and crystals.

## 12. Geochemical Models

In constructing a picture from data on the earth, several sources of information have been utilized:
a) the directly accessible parts of the earth's crust, including the oceans,
b) geophysical properties of the deeper parts of crust, mantle and core (cf. Schmucker, 1969),
c) experimental behavior of substances under high pressure and temperature (cf. Ringwood, 1970),
d) models for the formation of our planet (cf. Ringwood, 1966),
e) information from meteorites (cf. Keil, 1969) and recently from the moon.

The classical geochemical model for processes in the crust (Goldschmidt, 1933; Correns, 1948) assumed that all the sediments and the elements contained in sea water were derived from weathering of igneous rocks. Goldschmidt cal-culated the amount of weathered igneous rock on the basis of equilibria. He arrived at figures demonstrating that 160 kg/cm² of igneous rocks weathered to give 638 m of sediments. Calculations of Horn and Adams (1966) using 65 elements were that 400 kg/cm² weathered igenous rock gave 1590 m of sedi-ments. Many authors used only few elements for calculations with the classical model (cf. Conway, 1943; Wickman, 1954; Goldberg and Arrhenius, 1958; Brotzen, 1966). It is important to make the proper weighting of the different components (rock and sediment types), participating in the cycle and to obtain reliable averages for the concentrations of the respective elements in each domain. Horn and Adams (1966), when establishing their model for 65 ele-ments, used a computer to calculate the mean of minimum and maximum concentrations (from literature) for each element. They used a modified Poldervaart (1955) classification for continental, shelf, hemipelagic and pelagic sediments and also took into consideration evaporites and pore solutions of the sediments.

While most of the elements could be balanced in the crust, Cl, S, Mn, Br, B, Pb, As, Mo, J, Se, Sb occurred in excess in sediments and ocean waters. For the seven elements underlined volcanic sources probably account for the additional input.

Turekian and Wedepohl (1961), Vinogradov (1962), Taylor (1964), Ronov and Yaroshevskiy (1969) and Wedepohl (1969) calculated the proportions of all or most elements in the composition of the earth's crust. There are few differences between their resulsts and those of Horn and Adams.

Wedepohl (1972) pointed out that many igneous rocks of granitic or granodioritic composition are not primary, but are products of sediment metamorphosis. Hence, much material from these igneous rocks is "cyclic".

Conditions of rock weathering, sediment formation and therefore geochemical processes as a whole were different in the geological past (Veizer, 1973): oxygen and carbon dioxide partial pressures were different and there were more basic rocks available for weathering. The Precambrian sediments of the Canadian and Scottish shields are discussed by J. G. Holland and Lambert (1973) in relation to the composition of the continental crust.

Can geochemical cycles be regarded as proceeding in closed systems, *e.g.* the upper crust? Now that we recognize seafloor spreading, plate tectonics, and thrusting of the oceanic crust under island arcs and continental margins, we have to consider how much of the marine sediment is carried tens of kilometers or more down into the deep crust or upper mantle and is thus, at least for a while, removed from the geochemical cycle of the upper crust.

This model also implies that geochemical cycles go down as far as the upper mantle. The extent and significance of this process has not yet been estimated.

## 13. Literature

Those who wish to inform themselves about geochemistry find a number of periodicals which are exclusively devoted to this subject:

Geochimica et Cosmochimica Acta. Published by the Geochemical Society of the USA. Oxford-London-New York-Paris: Pergamon Press.
Geochemistry International. Mostly translations from "Geokhimia" (published by the Academy of Science, USSR). Published by the Geochemical Society of the USA.
Chemical Geology. Amsterdam: Elsevier Publishing Company.
Journal of Geochemical Exploration. Amsterdam: Elsevier Publishing Company.
Geochemical Journal. Published by the Geochemical Society of Japan.
Chemie der Erde. Jena: VEB Gustav Fischer Verlag.

*Introductory books to be mentioned are:*

Masons, B.: Principles of geochemistry, third edition. New York: Wiley & Sons 1966
Krauskopf, V. B.: Introduction to geochemistry. New York: McGraw Hill Inc. 1967
Wedepohl, K. H.: Geochemie. Berlin: W. d. Gruyter und Co. 1967.

Saukow, A. A.: Geochemie. Berlin: VEB Verlag Technik 1953.
Stumm, W., Morgan, J.: Aquatic chemistry. New York-London-Sidney-Toronto: Wiley Interscience 1970
Garrels, R. M., Christ, C. L.: Solutions, minerals, equilibria. New York: Harper & Row 1965
Schroll, E.: Analytische Geochemie. Stuttgart: F. Enke 1973

*More recent handbooks and tables are:*

Roessler, H. J., Lange, H.: Geochemische Tabellen. Leipzig: VEB Deutscher Verlag für Grundstoffindustrie 1965. Geochemical tables 448 pp. Amsterdam: Elsevier Publ. Comp. 1973
Wedepohl, K. H. (ed.): Handbook of geochemistry. Berlin-Heidelberg-New York: Springer 1969
*The Encyclopedia of Earth Sciences.* No 4: Fairbridge, R. W.: The encyclopedia of geochemistry and environmental sciences, 1321 pp. New York 1972.

*Several series continuously published on geochemical problems:*

Physics and Chemistry of the Earth: Pergamon Press.
Data of Geochemistry: Professional Papers of the USGS International Series of Monographs in Earth Sciences: Pergamon Press.

*Other books:*

Vinogradov, A. P.: Chemistry of the earth's crust (2 volumes). Transl.: Isr. Progr. Scient. Transl. 1966/1967
*International Geochemical Congress,* Moscow 1971, (2 volumes) (Abstracts of Reports), 1023 pp.

# 14. References

Adams, J. S., Gasparini, P.: Gamma-ray spectrometry of rocks. In: Methods in geochemistry and geophysics, 295 pp. Amsterdam: Elsevier Publ. 1970.
Ahrens, L. H.: The lognormal distribution of elements (1). Geochim. et Cosmochim. Acta *5*, 49–73 (1954).
Ahrens, L. H.: The lognormal distribution of elements (2). Geochim. et Cosmochim. Acta *6*, 121–131 (1954).
Ahrens, L. H.: Lognormal-type distributions (III). Geochim. et Cosmochim. Acta *11*, 205–212 (1957).
Ahrens, L. H.: Lognormal-type distributions in igneous rocks (IV). Geochim. et Cosmochim. Acta *27*, 333–343 (1963).
Ahrens, L. H.: Lognormal-type distributions in igneous rocks (V). Geochim. et Cosmochim. Acta *27*, 877–890 (1963).
Angino, E. E., Billings, G. K.: Geochemistry of subsurface brines. Symp. Proceed. Chem. Geol. *4*, 7–371 (1968).
Angino, E. E., Billings, G. K.: Atomic absorption spectrometry in geology. In: Methods in geochemistry and geophysics, Ser. 7, 191 pp. Amsterdam: Elsevier Publ. Co. 1972.
Banat, K., Förstner, U., Müller, G.: Schwermetalle in Sedimenten von Donau, Rhein, Ems, Weser und Elbe im Bereich der Bundesrepublik Deutschland. Naturwiss. *59*, 525–528 (1972).
Banat, K., Förstner, U., Müller, G.: Schwermetall-Anreicherungen in den Sedimenten wichtiger Flüsse im Bereich der Bundesrepublik Deutschland – eine Bestandsaufnahme. Vorläufiger Bericht; Laboratorium für Sedimentforschung, Universität Heidelberg; 230 pp. (1972).
Banno, S., Matsui, Y.: On the formulation of partition coefficients for trace elements distribution between minerals and magma. Chem. Geol. *11*, 1–15 (1973).

169

H. Puchelt

Barnes, H. L.: Geochemistry of hydrothermal ore deposits 670 pp. New York: Holt, Rinehart and Winston, Inc. 1972.

Barnes, H. L.: Deposition of Hydrothermal Ores. 24th Int. Geol. Congr., Montreal, Sect. 10, Geochemistry, 213 (1972).

Barth, T. F. W.: Theoretical petrology. 387 pp. New York: John Wiley and Sons 1952.

Barton Jr., P. B., Skinner, B. J.: Sulfide mineral stabilities. In: Geochemistry of hydrothermal ore deposits (ed. Barnes), p. 236–333. New York: Holt, Rinehart and Winston, Inc. 1967.

Berner, R. A.: Activity coefficients of bicarbonate, carbonate and calcium in seawater. Geochim. et Cosmochim. Acta 29, 947–965 (1965).

Bigeleisen, J.: Statistical mechanics of isotope effects on the thermodynamic properties of condensed systems. J. Chem. Phys. 34, 1485–1493 (1961).

Bigeleisen, J., Mayer, M. G.: Calculation of equilibrium constants for isotopic exchange reactions. J. Chem. Phys. 15, 261 (1947).

Bischof, G.: Lehrbuch der chemischen und physikalischen Geologie, II. Aufl. Bonn (1845–1854).

Bodu, R., Bouzigues, H., Morin, N., Pfeiffelmann, J. P.: Sur l'existence d'anomalies isotopiques rencontrees dans l'uranium du Gabon. C. R. Acad. Sc. Paris, Serie D, 275, 1731–1732 (1972).

Bolter, E., Turekian, K. K., Schutz, D. F.: The distribution of rubidium, cesium and barium in the oceans. Geochim. et Cosmochim. Acta 28, 1459–1466 (1964).

Bölviken, B.: A statistical approach to the problem of interpretation in geochemical prospecting, pp. 564–567. Toronto: The Canad. Inst. of Mining and Metallurgy, Spec. Vol. 11: Geochemical Exploration (1971).

Boström, K.: The problem of excess manganese in pelagic sediments. In: Researches in Geochemistry, Vol. 2 (ed. Ph. Abelson), pp. 421–452. New York: John Wiley and Sons Inc. 1967.

Bowen, R.: Paleotemperature analysis. In: Methods in Geochemistry and Geoph., Ser. 2, 265 pp. Amsterdam: Elsevier Publ. Comp. 1966.

Boyle, R. W., McGerrigle, J. I. (ed.): Geochemical exploration, 594 pp. Toronto: The Canad. Inst. of Mining and Metallurgy, Spec. Vol. 11 (1971).

Boyle, R. W., Shaw, D. M. (ed.): Exploration Geochemistry in Glaciated Terrains, 359–401. 24th Intern. Geol. Congr., Montreal, Sect. 10, Geochemistry (1972).

Braitsch, O.: Zur Geochemie des Broms in salinaren Sedimenten. Teil 1: Experimentelle Bestimmung der Br-Verteilung in verschiedenen natürlichen Salzsystemen. Geochim. et Cosmochim. Acta 27, 361–391 (1963).

Broecker, W. S., Gerhard, R. D., Ewing, M., Heezen, B. C.: Geochemistry and physics of ocean circulation. In: Oceanography (ed. M. Sears), Publ. 67. Washington D. C.: Am. Assoc. Adv. Sci. 1961.

Broecker, W. S., LI, Y. H., Cromwell, I.: Radium-226 and Radon-222 concentrations in atlantic and pacific oceans. Science 158, 1307 (1967).

Brotzen, O.: The average igneous rock and the geochemical balance. Geochim. et Cosmochim. Acta 30, 863–868 (1966).

Brunfelt, A. O., Steinnes, E.: Activation analysis in geochemistry and cosmochemistry, 468 pp. Oslo: Universitetsforlaget 1971.

Bryan, W. B., Finger, L. W., Chayes, F.: Estimating proportions in petrographic mixing equations by least-squares approximation. Science 163, 926–927 (1969).

Burns, R. G., Fyfe, W. S.: Crystal-field theory and the geochemistry of transition elements, In: Researches in Geochemistry, Vol. 2 (ed. Ph. Abelson), pp. 259–285. New York: John Wiley and Sons 1967.

Cannon, H. L., Hopps, H. C.: Environmental geochemistry in health and disease. Memoir 123, Geolog. Soc. Amer., 230 pp. (1971).

Chatters, R.: Radiocarbon and tritium dating. Washington, D. C.: Pulman 1965.

Chow, T. J., Goldberg, E. D.: On the marine geochemistry of barium. Geochim. et Cosmochim. Acta 20, 192–198 (1960).

170

Clarke, F. W.: The data of geochemistry. U. S. Geol. Surv. Bull. *770* (1924), 841 pp.

Conway, E. J.: Mean geochemical data in relation to oceanic evolution. Royal Irish Acad. Pr. 48, Sed. B. *8*, 119–159 (1942).

Conway, E. J.: The chemical evolution of the ocean. Royal Irish Acad. Pr. 48, Sed. B. *9*, 161–212 (1943).

Conway, E. J.: Mean losses of Na, Ca, etc. in an weathering cycle and potassium removal from the ocean. Am Jour. Sci. *243*, 583–605 (1945).

Correns, C. W.: Die geochemische Bilanz. Naturwiss. *35*, 7–12 (1948).

Correns, C. W.: The Discovery of the chemical elements. The history of geochemistry. Definitions of Geochemistry. Chapter I, Handbook of geochemistry, Vol. 1 (ed. Wedepohl) 1–II. Berlin–Heidelberg–New York: Springer 1969.

Craig, H.: Abyssal carbon and radiocarbon in the Pacific. J. Geophys. Res. *74*, 5491 (1969).

Craig, H.: The Geosecs program: 1970–1971. Earth and Planet. Science Letters *16*, 47–49 (1972).

Craig, H., Gordon, L. T.: Isotopic oceanography: deuterium and oxygen 18 variations in the ocean and marine atmosphere. Narragansett Mar. Lab. Oc. Publ. No. *3*, 277–374 (1965).

Craig, H., Miller, S. L., Wasserburg, G. J.: Isotopic and cosmic chemistry. 553 pp. Amsterdam: North-Holland Publ. Comp. 1964.

Dale, I. M., Henderson, P.: The partition of transition elements in phenocryst-bearing basalts and the implications about melt structure. 24th Intern. Geol. Congr., Montreal, Sect. 10, Geochemistry, pp. 105–111 (1972).

Dalrymple, G. B., Lanphere, M. A.: Potassium-Argon Dating, 258 pp. San Francisco: Freeman and co. 1969.

Danielson, A., Lundgren, F., Sundquist, G.: The tape machine. I, II, III. Spectrochim. Acta, 122–137 (1959).

Degens, E. T.: Geochemistry of sediments. 342 pp. New York: Prentice Hall, Inc. 1965.

Dietrich, G.: General oceanography, 588 pp. New York: Interscience Publ. 1963.

Dittmar, W.: Report on the scientific results on the voyage of H. M. S. "Challenger", 1873–1876. Physics and Chemistry *1* (1884).

Doe, B. R.: Lead isotopes. In: Minerals, rocks and inorganic materials, Vol. 3, 137 pp. Berlin–Heidelberg–New York: Springer 1970.

Durum, W. H., Heidel, S. G., Tison, L. I.: World-wide runoff of dissolved solids. Intern. Assoc. of Scientific Hydrology, General Assembly of Helsinki, pp. 618–634 (1960).

Edmond, J. M., Gieskes, J. M. T. M.: On the calculation of the degree of saturation of seawater with respect to calcium carbonate under in situ conditions. Geochim. et Cosmochim. Acta *34*, 1261–1291 (1970).

Ellis, A. J.: The chemistry of some explored geothermal systems. In: geochemistry of hydrothermal ore deposits (ed. Barnes), pp. 465–514. New York: Holt, Rinehart and Winston 1967.

Engelhardt, W. v.: Die Bildung von Sedimenten und Sedimentgesteinen, in press. Stuttgart: Schweitzerbarth-Verlag 1973.

Epstein, S., Taylor Jr., H. P.: Variations of $^{18}O/^{16}O$ in minerals and rocks. In: Researches in geochemistry (ed. Ph. Abelson), pp. 29–62. New York: John Wiley and Sons Inc. 1967.

Eremenko, N. A., Pankina, R. G.: On the evolution of ocean salt composition on the basis of the $^{32}S/^{34}S$ ratio in sulphate sulphur. 24th Intern. Geol. Congr., Montreal, Sect. 10, Geochemistry, pp. 291–295 (1972).
1. European Colloquium of Geochronology. Annales de la Societe Geologique de Belgique *94*, 95–137 (1971).
2. European Colloquium of Geochronology. Fortsch. Miner. *50*, Beiheft 3, 1–155 (1973).

Fairbairn, H. W., Schlecht, W. G., Stevens, R. E., Dennen, W. H., Ahrens, L. H., Chayes, F.: A co-operative investigation of precision and accuracy in chemical, spectrochemical and modal analysis of silicate rocks. U. S. Geol. Surv. Bull. *980* (1951).

Faure, G., Powell, I. L.: Strontium isotope geology. Minerals, rocks and inorganic materials, Vol. 5, 188 pp. Berlin–Heidelberg–New York: Springer 1972.

Fersam, A. Y.: Geochemical and mineralogical methods of prospecting. Chap. IV, Akad. Nauk. S. S. R., 164–238 (1939). Translation by Hartsock and Pierce, U. S. G. S. Circ. *127* (1952).

Fisher, F. H.: Effect of pressure on sulfate ion association and ultrasonic absorption in sea water. Geochim. et Cosmochim. Acta *36*, 99–101 (1972).

Flanagan, F. J.: 1972 values for international geochemical reference samples. Geochim. et Cosmochim. Acta *37*, 1189–1200 (1973).

Gaddum, J. H.: Lognormal Distributions. Nature No. 3964, Vol. *156*, 463–466 (1945).

Garrels, R. M., Christ, C. L.: Solutions, minerals and equilibria, 450 pp. New York: Harper & Row Publ. 1965.

Garrels, R. M., Mackenzie, F. T.: Evolution of sedimentary rocks: a geochemical approach, 397 pp. New York: Norton 1971.

Gast, P. W.: Isotope geochemistry of volcanic rocks in basalts. In: The Poldervart Treatise on Rocks of Basaltic Composition (ed. H. H. Hess and A. Poldervart) Vol. *1*, pp. 325–358. New York: Interscience Publ. 1967.

Goldberg, E. D.: Chemical and mineralogical aspects of deep sea sediments. Chap. 8 in Physics and Chemistry of the Earth, Vol. *IV* (ed. L. H. Ahrens), pp. 281–302. Oxford: Pergamon Press 1961.

Goldberg, E. D.: Minor elements in sea water. In: Chemical Oceanography (ed. I. P. Riley and G. Skirrow), Vol. *1*, pp. 163–196. London: Academic Press 1965.

Goldberg, E. D., Arrhenius, G. O. S.: Chemistry of pacific pelagic sediments. Geochim. et Cosmochim. Acta *13*, 153–212 (1958).

Goldschmidt, V. M.: Grundlagen der quantitativen Geochemie. Fortschr. Miner., Krist., Petrog. *17*, 112–156 (1933).

Goldschmidt, V. M.: The principles of distribution of chemical elements in minerals and rocks. J. Chem. Soc., 655–672 (1937).

Goldschmidt, V. M.: Geochemistry (ed. Alex Muir), 730 pp. Oxford: Clarendon Press 1954.

Graf, D.: Geochemistry of carbonate sediments and sedimentary carbonate rocks. I, II, III. Illinois State Geol. Surv. Circ. 297, 298, 301 (1960).

Grey, D. C.: Radiocarbon in the earth system. In: Minerals, rocks and inorganic materials (eds. W. v. Engelhardt, T. Hahn, R. Roy, P. J. Wyllie). Berlin–Heidelberg–New York: Springer, in press.

Häkli, T., Wright, T. L.: The fractionation of nickel between olivin and augite as a geothermometer. Geochim. et Cosmochim. Acta *31*, 877–884 (1967).

Hamilton, E. I.: Applied Geochronology, 267 pp. London–New York: Academic Press 1965.

Hawkes, H. E., Webb, J. S.: Geochemistry in mineral exploration. Harper's Geoscience Series (ed. Croneis), 415 pp. New York: Harper & Row Publ. 1962.

Hemphill, D. D. (ed.): Trace substances in environmental health. V. University of Missouri Press (1972).

Helgeson, H. C.: A chemical and thermodynamic model of ore deposition in hydrothermal systems. Mineral. Soc. Amer. Spec. Pap. *3*, 155–186 (197).

Helgeson, H. C.: Thermodynamic Constraints on Mass Transfer in Hydrothermal Systems at High Temperatures and Pressures. 24th Intern. Geol. Congr., Montreal, Sect. 10, Geochemistry, P. 230 (1972).

Hirt, B., Tilton, G. R., Herr, E., Hoffmeister, W.: The half-life of Re[187]. In: Earth Science and Meteoritics (ed. J. Geiss and D. Goldberg), pp. 273–280. Amsterdam: North-Holland Publ. Comp. 1963.

Hoefs, J.: Applied stable isotope geochemistry. Minerals, rocks and inorganic materials, Vol. *9*. Berlin–Heidelberg–New York: Springer 1973.

Holland, H. D.: The geologic history of sea water – an attempt to solve the problem. Geochim. et Cosmochim. Acta *36*, 637–651 (1972).

Holland, I. G., Lambert, R. St. J.: Major element chemical composition of shields and the continental crust. Geochim. et Cosmochim. Acta *36*, 673–683 (1972).

Holser, W. T., Kaplan, I. R.: Isotope geochemistry of sedimentary sulfates. Chem. Geol. *1*, 93–135 (1966).

Horn, M. K., Adams, J. A. S.: Computer-derived geochemical balances and element abundances. Geochim. et Cosmochim. Acta *30*, 279–297 (1966).

IAEA Panel: Nuclear techniques for mineral exploration and exploitation, 187 pp. Vienna: IAEA 1971.

Iiyama, J. T.: Fixation des elements alcalinoterreux Ba, Sr et Ca dans les feldspaths – etude experimentale. 24th Intern. Geol. Congr., Montreal, Sect. 10, Geochemistry, pp. 122–130 (1972).

Jäger, E.: Radiometrische Altersbestimmung in der Erforschung metamorpher Prozesse. Fortschr. Miner. *47*, 77–83 (1970).

Jasmund, K., Seck, H. A.: Partition of Elements in Coexisting Feldspars as Determined by Experiment and in Trachytic Rocks. 24th Intern. Geol. Congr., Montreal, Sect. 10, Geochemistry, pp. 78–84 (1972).

Jensen, M. L.: Biogeochemistry of sulfur isotopes, 193 pp. Proc. of a Nat. Science Found. Symp. New Haven, Conn.: Yale Univ. 1962.

Keil, K.: Meteorite composition. Chap. 4, Handbook of Geochemistry, Vol. *1*, (ed. Wedepohl), pp. 78–115. Berlin–Heidelberg–New York: Springer 1969.

Kharkar, D. P., Lal, D., Somayajulu, B. L. K.: Investigations in marine environments using radioisotopes produced by cosmic rays. In: Radioactive Dating, P. 175. Vienna: IAEA 1963.

Krauskopf, K. B.: Factors controlling the concentration of thirteen rare metals in sea water. Geochim. et Cosmochim. Acta *9*, 1–32 (1956).

Kretz, R.: Theory of chemical exchange reactions in crystalline rocks. 24th Intern. Geol. Congr., Montreal, Sect. 10, Geochemistry, pp. 67–77 (1972).

Kudo, A. M., Weill, D. F.: An igneous plagioclase thermometer. Contrib. Miner. Petrol. *25*, 52–65 (1970).

Kullerud, G.: Sulfide phase relations. Mineral. Soc. Amer. Spec. Pap. *3*, 199–210 (1970).

Kulms, M., Siemes, H.: Die Anwendung der „Trend-Surface-Analysis" im Rahmen geochemischer Untersuchungsarbeiten. Erzmetall, *23*, 371–378 (1970).

Kuroda, P. K.: On the nuclear physical stability of the uranium minerals, and: On the infinite multiplication constant and the age of uranium minerals. J. Chem. Phys. *25*, 781 and 1295 (1956).

Lal, D.: Characteristics of large scale oceanic circulation as derived from the distribution of radioactive elements. Morning Review Lectures, Int. Oceanogr. Congr., Moscow 1966, UNESCO 1969.

Libby, W. F.: Tritium in hydrology and meteorology. In: Researches in Geochemistry (ed. Ph. Abelson), pp. 151–168. New York: John Wiley & Sons 1959.

Livingstone, D. A.: Chemical composition of rivers and lakes. U. S. Geol. Surv. Prof. Pap. *440–G*, 64 pp. (1963).

McIntyre, W. L.: Trace element partition coefficients; a review of theory and applications to geology. Geochim. et Cosmochim. Acta *27*, 1209–1264 (1963).

Meier, H., Zimmermann, E., Albrecht, W., Bösche, D., Hecker, W., Menge, P., Unger, E., Zeitler, G.: On the existence of Promethium in nature. 24th Intern. Geol. Congr., Montreal, Sect. 10, Geochemistry, pp. 186–192 (1972).

Manson, V.: Geochemistry of Basaltic Rocks: Major Elements. In: Vol. *1*, Basalts; Poldervart Treatise on Rocks of Basaltic Composition (ed. Hess), pp. 215–270. New York: Interscience Publ. 1967.

Middleton, G. V.: Statistical Interference in Geochemistry. In: Studies in Analytical Geochemistry (ed. D. M. Shaw), pp. 124–139. The Royal Soc. of Canada, Spec. Publ. No. *6*, University of Toronto Press (1963).

Millero, F. J.: The partial molal volumes of ions in sea water. Limnol. Oceanogr. *14*, 376–385 (1969).

Millero, F. J., Berner, R. A.: Effect of pressure on carbonate equilibria in seawater. Geochim. et Cosmochim. Acta *36*, 92–98 (1972).

Moxham, R. M., Senftle, F. E., Boynton, G. R.: Borehole Activation Analysis by Delayed and Capture Gamma Rays Using a $^{252}$Cf Neutron Source. Econ. Geol. *67*, 579–591 (1972).

Müller, G., Förstner, U.: Cadmium-Anreicherung in Neckar-Fischen. Naturwiss. *60*, 258 (1973).

Neuilly, M., Bussac, I., Frejacques, Nief, G., Vendryes, G., Yvon, I.: Sur l'existence dans un passe recute d'une reaction en chaine naturelle de fissions, dans le gisement d'uranium d'Oklo (Gabon). C. R. Acad. Sc. Paris *275*, Serie D. 1847–1849 (1972).

Nichol, J., Garret, R. G., Webb, J. S.: The role of some statistical and mathematical methods in the interpretation of regional geochemical data. Econ. Geol. *64*, 204–220 (1969).

Nielsen, H.: Sulphur isotopes and the formation of evaporite deposits. In: Geology of saline deposits. Proc. Hannover Symp. 1968; UNESCO 1972, pp. 91–102 (1972).

Nielsen, H.: Stable sulphur isotopes. Chap. 16 B. *Vol. II, 4* Handbook of Geochemistry (ed. Wedepohl), in press. Berlin–Heidelberg–New York: Springer 1974.

Nielsen, H., Ricke, W.: Schwefel-Isotopenverhältnisse von Evaporiten aus Deutschland. Geochim. et Cosmochim. Acta *28*, 577–591 (1964).

Nordeng, S. C.: Application of trend surface analysis to semi-quantitative geochemical data. In: Short Course and Symposium on Computers and Computer Applications in Mining and Exploration, Tuscon, Ariz., Vol. *1*; College of Mines, The University of Arizona (1965).

Owen, B. B., Brinkley Jr., S. R.: Calculations on the effect of pressure upon ionic equilibria in pure water and salt solutions. Chem. Rev. *29*, 461–474 (1941).

Pavicevic, M., Ramdohr, P., El Goresy, A.: Electron microprobe investigations of the oxidationstates of iron and titanium in ilmenite in apollo 11, apollo 12, and apollo 14 cristallin rocks. Proc. 3. Lunar Science Conf. Suppl. 3 (Geochim. et Cosmochim. Acta) Vol. 1, 295–303. M.I.T. Press 1972.

Philpotts, J. A.: Redox estimation from a calculation of $Eu^{2+}$ and $Eu^{3+}$ concentrations in natural phases. Earth and planet. Science Letters *9*, 257–268 (1970).

Philpotts, J. A., Schnetzler, C. C.: Phenocryst-matrix partition coefficients for K, Rb, Sr and Ba with application to anorthosite and basalt genesis. Geochim. et Cosmochim. Acta *34*, 307–322 (1970).

Philpotts, J. A., Schnetzler, C. C.: Large Trace Cation Partitioning in Igneous Processes. 24th Intern. Geol. Congr., Montreal, Sect. 10, Geochemistry, pp. 51–59 (1972).

Poldervaart, A.: Chemistry of the earth's crust (ed. A. Poldervaart). Geol. Soc. Amer. Spec. Pap. *62*, 119–184 (1955).

Prinz, M.: Geochemistry of basaltic rocks: Trace elements. In: Vol. *1* Basalts; Poldervaart Treatise on rocks of basaltic composition (ed. Hess), pp. 271–324. New York: Interscience Publ. 1967.

Rankama, K.: Isotope geology, 535 pp. London: Pergamon Press 1954.

Rankama, K.: Progress in isotope geology, 705 pp. New York: Interscience Publ. 1963.

Rankama, K., Sahama, Th. G.: Geochemistry, 912 pp. Chicago University Press (1950).

Ringwood, A. E.: Chemical evolution of the terrestrial planets. Geochim. et Cosmochim. Acta *30*, 41–104 (1966).

Ringwood, A. E.: Phase transformations and the constitution of the mantle. Phys. Earth. Planet. Interiors *3*, 109–155 (1970).

Roedder, E.: Results and Significance of Recent Fluid Inclusion Studies in Ore Deposits. 24th Intern. Geol. Congr., Montreal, Sect. 10, Geochemistry, p. 231 (1972).

Ronoy, A. B., Yaroshevskiy, A. A.: Chemical composition of the earth's crust. In: The earth's crust and upper mantle. Geophys. Monogr. *13* (ed. P. J. Hart), pp. 37–57. Amer. Geophys. Union (1969).

Rubey, W. W.: The geologic history of sea water – an attempt to state the problem. Bull. Geol. Soc. Amer. *62*, 1111–1148 (1950).

Sakai, H.: Fractionation of sulphur isotopes in nature. Geochim. et Cosmochim. Acta *12*, 150 (1957).

Sakai, H.: Isotopic properties of sulfur compounds in hydrothermal processes. Geochemical J. 2, 29–49 (1968).

Sasaki, A.: Isotope geothermometry. Kozan Chishitsu 21 (5), 378–393 (Jap.) (1971).

Schaeffer, O. A., Zähringer, J.: Potassium argon dating, 234 pp. Berlin–Heidelberg–New York: Springer 1966.

Schmucker, U.: Geophysical aspects of structure and composition of the earth and the earth's Crust. Chap. 6, Vol. 1 Handbook of Geochemistry (ed. Wedepohl), pp. 134–226. Berlin–Heidelberg–New York: Springer 1969.

Schock, H. H., Puchelt, H.: Rubidium and cesium distribution in salt minerals. I. Experimental investigations. Geochim. et Cosmochim. Acta 35, 307–317 (1971).

Schoeller, H.: Geochimique des eaux souterraines. Application aux eaux des gisements de petrole. Rev. Inst. Franc. Petrole 10 (1955).

Schroll, E.: Analytical Geochemistry, in press. Stuttgart: F. Enke-Verlag 1973.

Schutz, D. F., Turekian, K. K.: The distribution of cobalt, nickel and silver in ocean water profiles around pacific antarctica. J. Geophys. Res. 70, 5519–5528 (1965 b).

Schwarcz, H. P.: The stable isotopes of carbon. Chap. 6 B I, Vol. II, 1 Handbook of Geochemistry (ed. Wedepohl). Berlin–Heidelberg–New York: Springer 1969.

Senftle, F. E., Wiggins, P. F., Duffey, D., Philbin, P.: Nickel exploration by neutron capture gamma rays. Econ. Geol. 66, 583–590 (1971).

Shannon, R. D., Prewitt, C. T.: Effective ionic radii in oxides and fluoride. Acta Cryst. B 25, 925–946 (1969).

Shapiro, L., Brannock, W. W.: Rapid analysis of silicate, carbonate and phosphate rocks. U. S. Geol. Surv. Bull. 1144-A, 56 pp. (1962).

Shaw, D. M.: Element distribution laws in geochemistry. Geochim. et Cosmochim. Acta 23, 116–134 (1961).

Shaw, D. M.: Evaluation of Data. Chap. 11, Vol. 1 Handbook of Geochemistry (ed. Wedepohl), pp. 324–375. Berlin–Heidelberg–New York: Springer 1969.

Shaw, D. M., Bankier, J. D.: Statistical methods applied to geochemistry. Geochim. et Cosmochim. Acta 5, 111–123 (1954).

Smales, A. A., Wager, L. R. (ed.): Methods in geochemistry, 465 pp. London: Interscience Publ. 1960.

Somayajulu, B. L. K., Lal, D., Craig, H.: Silicon-32 profiles in the south pacific. Earth and Planet. Sc. Letters 18, 181–188 (1973).

Steinberg, M.: Isotope-ratio method for tracing atmospheric sulfur pollutants. Power Generation Environ. Change Symp. (Proc.), publ. 1971, 302 – 16 E, Cambridge, Mass. USA. (1969).

Stumm, W., Morgan, I. I.: Aquatic chemistry, 583 pp. New York: Wiley-Interscience 1970.

Sverdrup, H. U., Johnson, M. W., Fleming, R. H.: The oceans, 1087 pp. New York: Prentice Hall Inc. 1942.

Taylor, H. P.: Oxygen and hydrogen isotopes in petrology. In: Minerals, rocks and inorganic materials, in press. Berlin–Heidelberg–New York: Springer 1973.

Taylor, S. R.: Abundance of chemical elements in the continental crust: a new table. Geochim. et Cosmochim. Acta 28, 1273–1285 (1964).

Taylor, S. R.: Geochemistry of andesites. In: Origin and Distribution of Elements (ed. L. H. Ahrens), pp. 559–584. Pergamon Press 1968. Oxford: Pergamon Press 1968.

Thiergärtner, H.: Grundprobleme der statistischen Behandlung geochemischer Daten. Freib. Forsch.-H. C 237, 99 pp. Leipzig (1968).

Thode, H. G.: Sulfur isotope geochemistry and fractionation between coexisting sulfide minerals. Mineral. Soc. Amer. Spec. Pap. 3, 133–144 (1970).

Turekian, K. K.: Trace elements in sea water and other natural waters. Annuat Report AEC Contract AT (30–1) – 2912, Publ. Yale 2912–12.59 pp. (1966 a).

Turekian, K. K., Johnson, D. G.: The barium distribution in sea water. Geochim. et Cosmochim. Acta 30, 1153–1174 (1966 b).

Turekian, K. K., Schutz, D. F.: The investigation of the geographical and vertical distribution of several trace elements in sea water using neutron activation analysis. Geochim. et Cosmochim. Acta 29, 259–313 (1965 a).

Turekian, K. K., Wedepohl, K. H.: Distribution of the elements in some major units of the earth's crust. Geol. Soc. Amer. Bull. *72*, 175–192 (1932).

Urey, H. C., Brickwedde, F. G., Murphy, G. M.: An isotope of hydrogen of mass 2 and its concentration (abstract) Phys. Rev. *39*, 864 (1932).

Veizer, J.: Sedimentation in geologic history: Recycling vs. evolution or recycling with evolution. Contr. Mineral. Petrol. *38*, 261–278 (1973).

Vinogradov, A. P.: Search for ore deposits by means of plants and soils. Akad. Nauk. S.S.R., Trudy Biogeokhim. Lab., V. *10*, 3–27 (1954).

Vinogradov, A. P.: Average contents of chemical elements in the Earth's crust. Geokhimiya, 641–664 (1962).

Vinogradov, A. P. (ed.): Chemistry of the earth's crust. Vol. 1 (1966) 458 pp., Vol. 2 (1967) 705 pp. (1966/1967). Israel Prog. Sci. Transl., Jerusalem.

Vinogradov, V. I.: Isotopic composition of sulfur as a factor indicating constancy of its cyclic circulation in time. Chem. Geol. *10* (1), 99–106 (1972).

Vlasov, K. A.: Geochemistry and mineralogy of rare elements and genetic types of their deposits. I. Geochemistry of rare elements. Nauk Publishing House, Moscow 1964; Israel Progr. Scient. Transl., 1966.

Vogg, H.: personal communication (1973).

Volborth, A.: Elemental analysis in geochemistry. Part A: Major elements, 373 pp. Amsterdam: Elsevier Publ. Comp. 1969.

Weber, I. N.: Bibliography of stable isotopes of carbon and oxygen. Pennsylvania State University, State College, Pa. USA. (1967).

Wedepohl, K. H.: Die Zusammensetzung der Erdkruste. Fortschr. Miner. *46*, 145–174 (1969).

Wedepohl, K. H.: Geochemische Bilanzen. Akad. Wiss. Lit., Mainz; Abh. Math. Nat. Kl., No. *2*, 27–42 (1972).

Wickman, F. E.: The "total" amount of sediments and the composition of the "average igneous rock". Geochim. et Cosmochim. Acta *5*, 97–110 (1954).

White, D. E.: Mercury and base-metal deposits with associated thermal and mineral waters. In: Geochemistry of Hydrothermal Ore Deposits (ed. Barnes), pp. 575–631. New York: Holt, Rinehart and Winston 1967.

White, D. E., Hem, J. D., Waring, G. A.: Chemical composition of subsurface waters. U.S. Geol. Surv. Prof. Pap. *440–F*, 1–67 (1963).

Wlotzka, F.: Nitrogen isotopes. In: Handbook of geochemistry (ed. Wedepohl), Vol. II/3, Chap. 7 B. Berlin–Heidelberg–New York: Springer 1972.

York, D., Farquhar, R. M.: The earth's age and geochronology, 178 pp. Oxford: Pergamon Press, 1972.

Zemann, J.: Crystal chemistry. In: Handbook of geochemistry (ed. Wedepohl), Vol. I, Chap. 2, pp. 12–76. Berlin–Heidelberg–New York: Springer 1969.

Received June 4, 1973

A critical presentation, by approximately 70 specialists, of important facts about the distribution of the chemical elements and their isotopes in the earth and the cosmos

**Executive Editor:**
K. H. Wedepohl

**Editorial Board:**
C. W. Correns
D. M. Shaw
K. K. Turekian
J. Zemann

# Handbook of Geochemistry

The work consists of two volumes, published in installments.
A subscription price is applicable on orders for the complete handbook, valid until the last install-ment is published.
Each installment is available separately at list price.
One installment probably remains to be published.

The Handbook of Geochemistry offers a critical selection of important facts about the distribution of the chemical elements and their isotopes in the earth and the cosmos. Approximately 70 specialist authors have made this selection from the flood of information which resulted from improved investigative methods. The data are set out in the main part of the Hand-book (Vol. II) in tables and diagrams as an integral part of extensive dis-cussions on abundance, distribution, and behavior of the elements. As this book clearly shows, geochemistry and cosmochemistry are intimately linked with a number of other disciplines.
With the exception of the introductory part (Vol. I), the work is arranged according to the atomic numbers of the elements, each chapter being organized in the same way. Thus, the reader will find such frequently needed information as the crystal chemical properties of an element or its occurrence in meteorites or metamorphic rocks under the same section in each of the 65 chapters. The loose-leaf system enables the contributions to be published in random order, regardless of their position in the book, and revisions to be made as desired.

Published 1972

Previously published

## Vol. II/3

Third Installment
With 161 figures
IV, 845 pages. 1972
Loose-leaf binder
DM 258,–; US $105.80
For subscribers to the complete Handbook
DM 206,40; US $84.70

Prices are subject to change without notice

## Vol. I

With 60 figures. XV, 442 pages. 1969

## Vol. II/1

With 172 figures. X, 586 pages. 1969
Loose-leaf binder. Vols. I and II/1 comprise the first installment and are not sold separately.
Boxed DM 224,–; US $91.90
For subscribers to the complete Handbook
DM 179,20; US $73.50

## Vol. II/2

Second Installment. With 105 figures
IV, 667 pages. 1970. Loose-leaf binder DM 212,–; US $87.–
For subscribers to the complete Handbook
DM 169,60; US $69.60

The fourth installment (in prepara-tion) will complete the handbook

■ Prospectus on request

 **Springer-Verlag Berlin · Heidelberg · New York**
München · London · Paris · Sydney · Tokyo · Wien

# Physics and Chemistry in Space

A series of monographs written and published to serve the student, the teacher and the researcher with a clear and concise presentation of up-to-date topics of space exploration.

Edited by J. G. Roederer, Denver, Colo.

Editorial Board: H. Elsässer, Heidelberg; G. Elwert, Tübingen; L. G. Jacchia, Cambridge, Mass.; J. A. Jacobs, Edmonton, Alta.; N. F. Ness, Greenbelt; Md.; W. Riedler, Graz

## Volume 5
## Hundhausen: Coronal Expansion and Solar Wind

By A. J. Hundhausen, High Altitude Observatory, National Center for Atmospheric Research, Boulder, Colo. USA
With 101 figs. XII, 238 pp. 1972
Cloth DM 68,—; US $27.90

The author gives a physical interpretation of basic solar wind phenomena, based on a synthesis of interplanetary observations and theoretical models of the coronal expansion.

Prices are subject to change without notice

**Springer-Verlag
Berlin
Heidelberg
New York**

München London Paris
Sydney Tokyo Wien

## Volume 1
## Jacobs: Geomagnetic Micropulsations

By J. A. Jacobs, Killam Memorial, Professor of Science, The University of Alberta, Edmonton, Canada
With 81 figs. VIII, 179 pp. 1970
Cloth DM 36,—; US $14.80

A detailed account both of the morphology of geomagnetic micropulsations and of the various theories that have been proposed to explain them.

## Volume 3
## Adler/Trombka: Geochemical Exploration of the Moon and Planets

By Dr. I. Adler, Senior Scientist, and Dr. J. I. Trombka, both: Goddard Space Flight Center, NASA, Greenbelt, Md, USA
With 129 figs. X, 243 pp. 1970
Cloth DM 58,—; US $23.80

A review of progress in the geochemical exploration of the Moon and planets and of future plans for lunar and planetary exploration.

## Volume 2
## Roederer: Dynamics of Geomagnetically Trapped Radiation

By J. G. Roederer, Professor of Physics, University of Denver, Denver, Colo., USA
With 94 figs. XIV, 166 pp. 1970
Cloth DM 36,—; US $14.80

A concise, systematic and up-to-date discussion of the basic dynamical processes governing the earth's radiation belts, with guidelines for quantitative applications of the theory.

## Volume 4
## Omholt: The Optical Aurora

By A. Omholt, Professor of Physics, Universitetet i Oslo, Fysisk Institutt, Blindern, Oslo, Norway
With 54 figs. XIII, 198 pp. 1971
Cloth DM 58,—; US $23.80

This book deals with the optical aurora, its occurrence and properties and the way it is produced by the primary electrons and protons. The auroral spectrum and its excitation is treated in great detail.

Prices are subject to change without notice